Visionary Scientist

Swedenborg Studies / No. 8
Monographs of the Swedenborg Foundation

Visionary Scientist

The Effects of Science and Philosophy on Swedenborg's Cosmology

....

Inge Jonsson

Swedenborg Foundation Publishers
West Chester, Pennsylvania

This work was originally published as
Emanuel Swedenborg by Twayne Publishers Inc., 1971.

Swedenborg Studies is a scholarly series published by the Swedenborg Foundation.
The primary purpose of the series is to make materials available for understanding
the life and thought of Emanuel Swedenborg (1688–1772) and the impact his
thought has had on others. The Foundation undertakes to publish original studies
and English translations of such studies and to republish primary sources
that are otherwise difficult to access. Proposals should be sent to:
Senior Editor, Swedenborg Studies, Swedenborg Foundation,
320 North Church Street, West Chester, Pennsylvania 19380.

Library of Congress Cataloging-in-Publication Data
Jonsson, Inge
Visionary scientist : the effects of science and philosophy on Swedenborg's cosmology / Inge Jonsson.
cm. — (Swedenborg studies : no. 8)
Rev. ed. of : Emanuel Swedenborg.
Includes bibliographical references and index.
ISBN 0–87785–192–1. — ISBN 0–87785–191–3 (pbk.)
1. Swedenborg, Emanuel, 1688–1772. I. Jonsson, Inge. Emanuel Swedenborg.
II. Title. III. Series.
BX8748.J64 1999
289'.4' 092–dc21 99–34378
CIP

Translated by Catherine Djurklou
Edited by Mary Lou Bertucci
Cover and interior designed by Sans Serif
Set in minion by Sans Serif, Inc., Saline, Michigan

Printed in the United States of America.

Contents

	Preface to First Edition	vii
	Preface to Second Edition	ix
	Chronology	xi
Chapter 1	The Age and the Man	3
Chapter 2	Swedenborg's Philosophy of Nature	21
Chapter 3	From Inorganic to Organic Nature	35
Chapter 4	Ontology and Psychology	69
Chapter 5	Universal Philosophy and the Theory of Correspondence	93
Chapter 6	The Religious Crisis	123
Chapter 7	The Secrets of Heaven	143
Chapter 8	The Lord and His Word	175
Chapter 9	The New Jerusalem in the World	195
	Notes and References	211
	Selected Bibliography	231
	Index	245

Preface to the first edition

It is certainly not surprising that a great deal has been written about Emanuel Swedenborg ever since the end of the eighteenth century. Great authors normally attract a host of critics. But Swedenborgian specialists comprise a rather unusual group in the sense of the words of the Master that he who is not with me is against me. Consequently, there probably is a need for a new study by someone who is neither convert nor maligner.

It was thus especially difficult not to accept the offer of the publishers of the Twayne World Authors Series to write about my remarkable compatriot—perhaps the most remarkable of them all—even though it would be preposterous to attempt to compress the entire story of Swedenborg's life and works into the limited space at my disposition.

My main intention has been to approach Swedenborg's writings from the standpoint of the history of ideas, to try to treat them as a rare fruit on the tree of European science. After an introductory biographical sketch, Chapter 2 is devoted to Swedenborg's philosophy of nature, with the emphasis on cosmology and the theory of particles. The third chapter discusses his shift from the mechanistic concept of nature to an organic philosophy, which means that the great volumes and the even greater aspirations of the years 1734–1745 are the focus of attention. Chapter 4 provides a more detailed examination of Swedenborg's attempts to force his way into the concealed domains of the soul with the assistance of some metaphysical doctrines. The fifth chapter analyzes in still greater detail what Swedenborg called a mathematical philosophy of universals and his doctrine of correspondence. The reason for such detailed attention is that the theory of correspondence is one of the concepts for which Swedenborg is most widely known and also because

I hope to report on certain new findings. In Chapter 6, "The Religious Crisis," I turn to Swedenborg's activities after 1745, when the gates to the spiritual world, to heaven and to hell, were opened for him; this section is primarily concerned with the *Journal of Dreams* and the superb drama of the creation *De Cultu et Amore Dei* (The Worship and Love of God). The seventh chapter cites a number of examples of the various types of visions encountered in his manuscripts and published works; it also attempts to convey a sense of his completely humanized spiritual world. Chapter 8 discusses Swedenborg's concept of God and his view of the Bible and other essential theological problems, and the last chapter traces some of the channels of his influence on Western literature up to our own times.

I wish to thank the Royal Swedish Academy of Science for economic support and for the use of their library resources.

Inge Jonsson
Stockholm
August 1969

Preface to the second edition

Since this book has been out of print for many years, it was difficult not to accept the invitation from Swedenborg Foundation Publishers to issue a revised edition of it. Returning to my text after thirty years has been a rather complicated experience, including quite a few temptations of substantial changes of some parts, but I have tried to confine myself to making only necessary corrections and additions. There has been very little time for systematic research work in later years because of other academic duties, so maybe my revision has unintentionally become too restrictive. Nevertheless, I hope that it may serve as an introduction to one of the most fascinating writers in Western literature.

I wish to convey my deep gratitude to the publishers for having given me this opportunity to return to old hunting grounds, and particularly to Mary Lou Bertucci for her work as editor.

Inge Jonsson
Stockholm
1999

Chronology

1688	Emanuel Swedberg born (January 29) in Stockholm, son of Regimental Chaplain Jesper Swedberg and Sara Behm
1692	Jesper Swedberg promoted to a theological professorship at Uppsala University.
1696	Sara Behm dies June 17
1699	Emanuel Swedberg matriculates at Uppsala University
1703	Jesper Swedberg and his second wife, Sara Bergia, leave Uppsala for Skara, where he has been appointed bishop
1709	Emanuel Swedberg graduates with a thesis in moral philosophy
1710–1715	Young Swedberg makes his first journey abroad to England, Holland, France, and Germany; publishes a collection of Latin poems in Greifswald
1716	Publishes *Daedalus Hyperboreus*, Sweden's first scientific journal, and is appointed extraordinary assessor of the Board of Mines
1719	Is ennobled and assumes the name Swedenborg
1721–1722	Goes to Holland and Germany, where he publishes *A Specimen of a Work on the Principles of Natural Philosophy* and *Miscellaneous Observations on Natural Things*
1733–1734	Goes to Germany and Bohemia to publish his *Philosophical and Mineralogical Works* and *The Infinite*
1735	Bishop Swedberg dies July 7
1736–1740	Travels to Holland, France, and Italy for research work in anatomy and physiology; publishes *The Economy of the Animal Kingdom* in Amsterdam
1743–1745	Swedenborg's fifth foreign journey, during which he experiences his religious crisis and publishes *The Animal Kingdom* and *The Worship and Love of God*
1747	Leaves the Board of Mines

1749–1756	Publishes eight great volumes of *Heavenly Secrets* in London
1758	Publishes five works in London: *Earths in the Universe, Heaven and Hell, The Last Judgment, The White Horse,* and *The New Jerusalem and Its Heavenly Doctrine*
1760	Johann August Ernesti attacks Swedenborg's doctrines in his *New Theological Library*
1763–1764	Publishes *Divine Love and Wisdom, Divine Providence, The Doctrine of the New Jerusalem concerning the Lord; concerning the Sacred Scripture; concerning Life; concerning Faith* and *The Last Judgment Continued* in Amsterdam
1766	Publishes *The Apocalypse Revealed* in Amsterdam; is attacked by Immanuel Kant in *Dreams of a Spirit-Seer*
1768	Heresy trial of two of Swedenborg's disciples begins in Gothenburg; Swedenborg publishes *Marital Love* in Amsterdam
1769	Publishes *Brief Exposition of the Doctrine of the New Church* and *The Intercourse between the Soul and the Body* in Amsterdam and London
1771	Publishes *The True Christian Religion* in Amsterdam
1772	Dies in London March 29

Visionary Scientist

1

The Age and the Man

The Intellectual Climate

T.S. Eliot coined the phrase "dissociation of sensibility" to define the impact on Western intellectual history of the scientific breakthrough of the seventeenth century. As a result of a tremendous expansion in knowledge combined with attacks on the closed medieval interpretation of the world, the traditional view of humankind and of the world was thoroughly shaken and in its place came confusion and anguish. In his agonized *The First Anniversary*, John Donne bewails the fact that the new science "calls all in doubt" and demolishes all harmony—"all coherence gone." Nature's great book was no longer open for all to read of the glory of the Creator, as the psalmist said, but was intelligible only to those who spoke its tongue, the language of mathematics. That gift, then as now, was reserved for the few, and the poet was rarely one of them. A rift appeared between the exact language of formulae and the symbolism of poetry; in our time this rift has become what many regard as a threatening abyss.

The "century of genius" produced a miraculous development in the formulation of mathematics, to the point that the most remote movements in the macrocosm could be explained by formulae similar to those that dealt with the behavior of everyday objects. To mathematical visionaries, it was not inconceivable that a conceptual algebra could be created that would replace exact mathematical operations for the fruitless metaphysical bickerings of the philosophical schools. Even though the many attempts to realize the dream of a perfect universal language had not been successful, it is easy to imagine how vitally important the dream might become to one who had experienced it with all the emotional intensity of youth. From this mathematical sensibility sprang the new rational systems of thought, which strove to incorporate the innumerable scientific discoveries into new syntheses.

It is impossible to say who made the most significant contributions to the intellectual life of the seventeenth century, but the answer is unequivocal in the environment that shaped Emanuel Swedenborg: René Descartes. Although he was by no means the first to attack Aristotelian university philosophy, Descartes's assault was the most destructive one. Using radical doubt as a methodological tool, he shattered the exquisite edifice erected by the Scholastics, persisting until he had reached what he regarded as an absolutely stable ground for thought: the existence of the thinking individual. It was upon this ground that Descartes' system was built, but it differed in principle from the lost unity in its dualistic character. The material world with extension as its attribute was regarded as utterly different from psychical reality in which thought rules, and there was no real communication between these two worlds. This immediately created the problem of explaining the actually experienced interaction between soul and body in man, which came to be one of the most actively discussed problems in the succeeding hundred years. It was his attempts to find a solution that brought Emanuel Swedenborg to the gates of the spiritual world.

Swedenborg received his academic training at the University of Uppsala during the first decade of the eighteenth century. Although Descartes had been at the court of Queen Christina (1632–1654) during the last four months of his life—his delicate health could not stand the winter climate of Stockholm—his philosophy did not win acceptance at the university until late in the seventeenth century and only after harrowing struggles. The same was true all over Europe, the quarrels of the learned concerning the atheistic consequences that many theologians believed Cartesianism would inevitably lead to. This opinion was also shared by Voltaire in his book on Newton's philosophy, issued as late as 1731, and La Mettrie honored Descartes as "the intellectual ancestor of *L'homme machine*."[1] In Sweden, the argument was terminated in 1689 by royal decree from the absolute monarch. The result, according to the most probable interpretation, was somewhat of a victory for Descartes's theological opponents: his principles could only be applied to the natural sciences and to philosophy, and any encroachment on theological domains would be repulsed.[2]

The battle about Descartes's philosophy is an indication of the powerful position of theology and the Lutheran clergy in Sweden in the days when the country was a great power, as well as of the hysterical suspicion of all innovations. But the distrust of the new science on the part of European theologians was not wholly unjustified. At the same time, the great majority of the natural scientists of the day—many of them truly devout men—were extremely anxious to save the main tenets of the Christian faith. The host of anatomists included such dedicated Catholic converts as the Danes Steno and Winslow, a mystic like Swammerdam, and scientists like Méry and Heister, for whom even the tissues of the body bore witness to the glory of God. For a number of geniuses, led by Isaac Newton, perusal of the book of nature found its self-evident complement in study of the revelations of the Scriptures, an attitude that caused many of them to penetrate the texts in their original versions and to contribute to their interpretation. The development of microscopic techniques revealed an unknown world, and from this encounter were born the most jubilant paeans to the glory of the Lord. In ingenious philosophical syntheses, Malebranche and Leibniz, inspired by Descartes, were able to combine the new science with profound religious experience.

But the intellectual climate became increasingly chilly for orthodox Christianity during the course of the eighteenth century. In the long run, skepticism and disbelief grew ever stronger in the wake of the new science. Enlightened European thought became ever less inclined to listen to a gospel that was often empty words with no power to influence clerical conduct. In France, the cradle of the

Enlightenment, anti-religious feeling exploded at the end of the century in excesses of hatred of Christianity and of the substitutional cult of reason. But the churches were attacked not only by radical elements in the Enlightenment. In Protestant territory, an emotional protest against orthodoxy developed toward the end of the seventeenth century, taking the form of a series of new sects. At first these sects were persecuted with the same harshness as in the century of the religious wars, but with the passing years, the demand of the Enlightenment for religious tolerance gained increasing acceptance. Radical pietistic groups showed signs of an unhealthy exaltation, a repugnant obsession with the mystery of the stigmata, and grossly superstitious forms of spiritualism.

Martin Lamm, the great Swedish scholar who wrote an excellent book on Swedenborg in 1915, has coined the phrase "romanticism of the Era of Enlightenment" to characterize the many expressions of emotional revolt against the worship of reason and the utilitarian ideology of that age. Lamm's term allows us to stress that no absolute change of climate occurred with the romantic breakthrough in the decades on either side of the turn of the century in 1800. From our perspective, it was more a change in nuance and a reaction against the ideals that had guided the previous generation. It meant, among other things, that the rationalism of the seventeenth century was reassessed and won new disciples. We may recall that the central organism concept of Romanticism was inspired by a reinterpretation of Leibniz, as a result of the publication in 1765 of his hitherto unknown *Nouveaux Essais sur l'Entendement Humain* and also that Spinoza—that feared and despised pantheist—was eagerly read and loved by the young Goethe, for example.

In many cases, however, these connections are difficult to discern. It is particularly hard in Swedenborg's case, due to his having assumed prophetic dimensions following his call in 1745. It may seem risky or presumptuous to evaluate according to usual historical categories a man who saw himself as an emissary of God and looked upon his activities as an interpreter of the Scriptures as the return of the Messiah. But an attempt at fitting him into the traditional pattern can find support in the religious–historical view that the assumption of a religious role is always dependent on the entire makeup of the individual, including his level of education. As we shall see later, the events of the mid-1740s did not mark a complete break in Swedenborg's development. It is true that he almost entirely gave up his research in the natural sciences; but as an interpreter of the Bible, he was able to use the same methods as before, namely, to annotate the texts and provide his personal interpretation. He retained more or less intact the philosophical position at which he had arrived before his call, and for those who do not adhere to the faith of his disciples, it is his vision of the conditions of mankind that are fascinating, not his biblical writings as such.

From a profane point of view, Swedenborg's theory of correspondence is of central importance. In his great essay on the poetry of Victor Hugo, written in 1861, Baudelaire praised Swedenborg for having taught us that "heaven is a Great Man: that everything, form, movement, quantity, color, odor, both in the spiritual world and in nature, is significant, reciprocal, converse, correspondent."[3] This may sound like an inexact variation of a Platonic interpretation of the universe, but it has its principal sources in the thought of the times. To prove this, considerable attention will have to be paid to much that is bizarre and obsolete in the beliefs and thinking of the seventeenth and eighteenth centuries; and it can be stressed at this point that Swedenborg is neither an easy nor an immediately appealing writer, not even if one translates his sometimes rather uninspired Latin. But thrilling experiences await those who have the patience to try to penetrate Swedenborg's environment. To the man himself, his decades of struggling with disparate tendencies in the knowledge of the time yielded assurance and spiritual peace within the framework of a highly original form of Christianity. Bearing in mind his point of departure, one can borrow Eliot's phrase and assert that Swedenborg personally overcame the "dissociation of sensibility" of the seventeenth century. This may be the reason that so many great poets—from Blake to Yeats and Borges, Balzac and Baudelaire to Valéry, Stagnelius to Strindberg and Ekelund, to name a few—were attracted to him. From the point of view of literary history, Swedenborg is certainly the most influential of Swedish authors.

The Man and His Work

A few words need to be said about the facts of Swedenborg's life. Because many people around the world have long regarded him as a prophet of the Lord, whose every word must be retained, both biographically significant documents and the great majority of his manuscripts have been published. American Swedenborgians, in particular, have made a magnificent contribution in this respect, especially through R.L. Tafel's three great volumes of documents and through almost twenty volumes of photostat copies of original manuscripts. Swedenborgian research owes a great debt of gratitude to the generosity of these enthusiasts, but it is a pity that, to a certain extent, the work of publication was done in a way that does not meet scholarly standards. The documents have been translated into English for obvious reasons, but this entailed some misunderstandings of the texts; and, when manuscripts were photographed, the enormous amount of material was edited in such a way that the most important parts from the point of view of historical research were often left out. To appreciate Swedenborg's erudition and to recognize his sources of inspiration, it is still necessary to consult the manuscripts, which have been kept in the library of the Royal Academy of Sciences in Stockholm since December 1772, when the permanent secretary paid the haulage cost for a chest containing some 20,000 handwritten pages. A tremendous amount of work still remains to be done before all of these manuscripts will have been analyzed.

It is nevertheless easy to select the basic biographical data from Tafel's collection of documents. It is an odd coincidence that the man who is perhaps the most widely known Swedish writer in the English-speaking world bore a name with a direct association to his mother country. However, the prefix actually derives from the ancestral estate, Sweden, in the neighborhood of the town of Falun in Dalecarlia. On his father's side, Swedenborg descended from a mining family, and the name of the estate became the family name in the generation before his, when his father took the name of Swedberg as a student.

Jesper Swedberg was born in 1656 and in the course of his long life attained the highest positions in the church, bishop in 1702 and doctor of theology in 1705. According to the custom of the day, Bishop Swedberg's wife and children were ennobled, after which the name was changed to the more lustrous "Swedenborg" in 1719. Jesper Swedberg has become somewhat of a symbol for the type of piety prevalent at the time and of the strong self-reliance and sense of power of the Swedish clergy. In his lively and colorful memoirs, he obeyed the Scriptures and did not hide his light under a bushel. Grandiose, blustering, self-assured, unquestioningly devout, and artlessly naive, an egocentric of colossal dimensions, but indubitably a man with a tremendous capacity for work and an unusual feeling for the demands of his high calling, manifested not least by his efforts to create a hymnal for the Swedish church—this is how one can summarize the impression made by his autobiography, which is one of the most vivid personal documents of the Sweden of a few centuries ago.

Swedenborg had his roots in Bergslagen, the mining district, on both his mother's and father's sides; and, even though the father had chosen a career in the clergy, it was natural for the son to become an official in the Royal Board of Mines. His maternal grandfather, Albrecht Behm, was a prosperous mine owner, who also bore the same title that his famous grandson would one day be given, namely, assessor of the Board of Mines, the government agency that supervised the important mining industry in the country. When in 1683 the penniless regimental pastor Jesper Swedberg married Sara Behm, he acquired a great deal of worldly goods into the bargain, and even though several siblings eventually arrived to share the Behm gold, it constituted together with other inheritances the financial basis for Swedenborg's activity as a writer; although he lived economically as a bachelor, his extensive travels and the publication of his vast scientific and theosophical works required considerable expenditures.

Swedenborg's mother died in 1696, when he was only eight years old. At the time, his father was a professor of theology at the University of Uppsala, where Emanuel was matriculated in 1699. Little is known of his studies and even less of his relationship with his domineering father: there was little interest in psychological analysis in those days, and Jesper Swedberg's memoirs have nothing revealing to say about his children. Swedberg remarried in 1697, but Emanuel appears to have reacted entirely positively to his stepmother, a wealthy widow named Sara Bergia. At the same time, it is reasonable to assume that the pressure from his father was very strong, and certain letters written when he was in his twenties seem to reveal mild reactions of protest. But these date from a much later period, and it should be recalled that Emanuel remained at Uppsala when his father moved to Skara as bishop in 1703. In the sensitive years of his puberty, he found a second home with his oldest sister and her husband, who could be seen as a second father, a spiritual father.

Swedenborg's brother-in-law was Erik Benzelius, Jr., one of the greatest figures in the intellectual history of Sweden. In his home, the young student came into early contact with a many-faceted scholarly ideal, which had its roots in solid knowledge of the classics but also included the latest natural sciences. Benzelius could describe the European centers of learning he had come to know during his studies abroad in the last three years of the seventeenth century, and Emanuel probably read the lively correspondence that Benzelius, as university librarian, maintained with the great scholars of the period. Benzelius could tell of his meetings with some of them, not least the brilliant Leibniz, on whom he had made a good impression, and he could point to the practicable results which could be reasonably anticipated from the scientific breakthrough of the seventeenth century. Benzelius himself worked eagerly at spreading the new knowledge; and, in the tragic autumn of the 1710 plague, he took the initiative for the creation of Sweden's first learned society, *Collegium Curiosorum*, the Society of the Enquiring, in Uppsala, a modest counterpart of the Royal Society in London and the Académie des Sciences in Paris.

The basis for the discussions in this society consisted in part of reports from two correspondents, Christopher Polhem and Emanuel Swedberg. The latter had completed his academic studies in 1709 with the customary master's thesis—it dealt with a few maxims from Roman ethics—and, in spite of war and pestilence, he succeeded in reaching the objective he had early set for himself, namely, a period of study abroad, such as both his father and his brother-in-law had enjoyed. His interests differed from theirs, however. Emanuel Swedberg traveled first to England in order to familiarize himself with the new natural sciences. In letters to Benzelius, he wrote that he had an "immoderate desire" for mathematics and its applications to astronomy and mechanics.

Before his departure, Emanuel had prepared himself by collecting material for an investigation in the advances of mathematics during the most recent centuries, and in the first of his interesting letters to Benzelius, he told of his daily readings of Newton. Writers on Swedenborg usually claim that the young scholar met Newton and listened to his discourses, although we have no proof of this.[4] On the other hand, he did meet great men in the scientific world, including Flamsteed, Halley, Chamberlain, and Woodward. In addition to learning practical mechanical skills from craftsmen with whom he was lodging, Swedberg's time was devoted to astronomical problems; and he was especially absorbed by one question which, if he could have found the answer, would have brought him fame and fortune—how to determine longitude with the help of the moon, an assignment announced by the Royal Society. The purpose was to make navigation at sea safer. The young Swede, however, like so many others, was unable to solve the problem. It seems that his disappointment was so bitter that it became a kind of trauma: he returned to the problem on several occasions, even as an aged visionary in a paper sent to the Academy of Sciences in Stockholm; unfortunately, it met with as little success as had all the previous ones.

After about three years in England, Emanuel Swedberg moved to Holland, where he received his first impression of the scholarly environment that had meant so much to Swedish culture and learning in earlier days and still continued to do so. In his letters, however, he dwelt mainly on the peace conference in Utrecht, which he observed first hand. His next stop was Paris, where he sought out the most renowned mathematicians in order to present his solution of the lunar longitude problem to scholars less prejudiced than—but obviously just as skeptical as—the English. His last letters from abroad were posted in the German Baltic provinces of Sweden, which were in the process of being lost in the final stage of the Great Northern War. These letters reveal that the war was beginning to be a reality to the traveler, although his dominant interests remained scientific and technological. In general, national tragedies occupied a surprisingly small place in this correspondence, even that part written at the headquarters of Charles XII, where Emanuel Swedberg served on several occasions.

Meanwhile, his scientific work could be justified militarily to the extent that some of his many projects would have radically increased his country's strength. The designs he brought back from abroad included sketches of submarines and airplanes, and history would no doubt have been changed had they been realized. They were items in a long series of more or less fantastic plans, a list that not seldom recalls contemporary satires on the technological planners known as *virtuosos.* But the planner himself had a live example close at hand, a man whom even before his voyage he had looked up to as an ideal: Christopher Polhem (1661–1751), self-taught mechanic and inventor, had himself spent several years on the continent and in England in the 1690s in order to learn the latest developments in the practical application of mechanics, and had made important contributions after his return, for example, in mining technology and the art of warfare.

It is quite clear that while abroad Swedberg regarded himself as following in Polhem's footsteps, and the superficial contact he had had with his ideal before his voyage was revived after his return in 1715 through the intermediary of Benzelius. He wrote a treatise on some of Polhem's inventions, which was later published in *Daedalus Hyperboreus,* which Swedenborg edited and which was Sweden's first scientific journal, the first of six issues of which was printed in 1716. Polhem obviously had a high opinion of his editor's mathematical knowledge and mechanical talent, and the periodical was filled almost entirely with descriptions of Polhem's inventions. Swedenborg, nevertheless, found room for some of his own articles, mainly on mathematical subjects.

Daedalus Hyperboreus caused Swedenborg much trouble and some financial worry, but when the periodical came to the attention of Charles XII at his headquarters in Lund, it opened the door to his appointment as *assessor extra ordinem* at the Royal Board of Mines, with a special assignment as assistant to Polhem. Despite considerable opposition, Swedenborg was given tenure as assessor some years after the death of the autocratic king and remained in the position until 1747, when he resigned in order to devote all his energies to his research in spiritual subjects.

Swedenborg's relationship to Polhem was not wholly without friction, and it appears from Polhem's letters as if he was occasionally somewhat suspicious of the learned "young Mr. Swedberg" and at times displeased with inexact reviews and descriptions in the periodical. But his mildly paranoid attitude scarcely went much below the surface, and it is probable that Polhem would happily have accepted his assistant as a son-in-law if his daughter Emerentia had been willing. That Polhem was of great importance to Swedenborg's scientific development is beyond all doubt. When Swedenborg strove to include ever greater domains, including psychical, within a mechanistic philosophy, he had many forerunners among natural scientists of the Cartesian tradition in its broadest sense, but Polhem was the scholar to whom he was closest during several years of great significance to his development.

In addition to *Daedalus Hyperboreus*, Swedenborg published a series of short papers during the 1710s and the beginning of the 1720s, mathematical, geological, and cosmological treatises, but also a collection of poems in Latin. His activities subsided in the new political situation after the death of Charles XII in 1718, and Swedenborg complained bitterly of the lack of understanding of the importance of research shown during these early days of the rule of the Estates. His publication of a number of small chemical, physical, and mineralogical essays in Latin during a stay in Holland and Germany at the beginning of the 1720s was followed by a long silence, although these essays received positive critiques in the highly esteemed journal *Acta eruditorum* in Leipzig. Until 1734 Swedenborg's writing was almost entirely confined to questions relating to his position at the Board of Mines, including the first technological assessment undertaken in Sweden (of a Newcomen steam engine), and to contemporary political and economical problems.[5]

In 1729 when he was elected a member of *Societas Literaria et Scientiarum* in Uppsala, an offshoot of the *Collegium Curiosorum* of the 1710s, Swedenborg explained his silence in letters by the circumstance that in the space of ten years he had "assembled all the facts which contribute to knowledge in *Metallicis et regno minerali*."[6] A few years later, the results of these efforts were incorporated in the three great folios, *Opera Philosophica et Mineralia*, with which the 46-year-old Emanuel Swedenborg appeared before the learned readers of Europe and which soon gained him a great reputation as a mineralogist. In a letter to Benzelius, written in 1736, the famous German orientalist Christopher Wolf called his brother-in-law a *vir optimus et sagax* with few contemporary equals in natural history.[7]

It is difficult to define the general impressions conveyed by the personal documents and scientific papers mentioned above, mainly because they are so unsensational. We find a man with an alert and agile intellect and an insatiable appetite for all the fare offered by modern science, who through fortunate circumstances found himself in the centers of European research during the period when a scientist acquires his methodological principles. The mathematical language that Galileo regarded as essential for the reading of the book of nature was at its most advanced stage in the England of Newton. Swedenborg early understood the decisive importance of mathematics to modern research and tried both to learn and to disseminate the latest findings. It has been said that he actually had no great mathematical gift, but that his knowledge after his return from abroad was vastly superior to that of his compatriots. There is no question that he retained his youthful enthusiasm for the abstract beauty of the world of mathematics even into his theosophical period, as will be shown in greater detail later on.

There is a hint of immaturity in the rather boastful tone of Emanuel Swedenborg's letters from abroad. The traveler was extremely eager to point out how clever and diligent he was, and his many more or less impracticable research projects fit into the same pattern. But it would be unfair to over-emphasize this element and to interpret it as a symptom of instability. Instead it reveals a normal lack of confidence, as well as high aspirations. The letters reflect very little of Jesper Swedberg's boundless self-confidence and nothing of the father's naive belief in miracles, but in general they make a sober impression: the level of egocentricity scarcely exceeds what can be considered normal in a young scholar.

The young Swedenborg was thus neither a bookworm without contact with reality nor a marked careerist. What is most striking—and in many ways impressive—is his appetite for knowledge in so many areas and his willingness to share what he learned. It is important to emphasize the practical streak in Swedenborg, since it is easy to think of him as a mild dreamer, a converser with angels even as a young man. The heritage from generations of mining men predisposed him for the tasks he was given as Polhem's assistant and assessor at the Board of Mines, and even as an old man in his theosophical period, he continued to be concerned with political and financial problems and with finding solutions for them. In addition, Swedenborg was one of the most widely traveled Swedes of the eighteenth century, even though he never left Europe.

Extant diaries from some of his voyages yield further information about the man Swedenborg. The more detailed of them cover the years 1733–1734 and 1736–1739, the first in Latin, the second, strangely enough, in Swedish (Swedenborg usually wrote even outlines and private notes in Latin). In Swedish literature, Carl von Linné's descriptions of his travels became classics through their spontaneous charm, ability to capture swift impressions and moods, pure epical joy, and naive observations; but Swedenborg, an older relative of Linné, never achieved the same heights with his accounts of his travels. Swedenborg never intended to publish them, but they have been printed posthumously both in original and in translation.[8] His journals are also much less personal, contain many useful facts, and are mainly remarkable for what we are inclined to regard as typical of the Enlightenment: a thirst for all kinds of factual information, a receptivity to foreign cultural environments, religious tolerance, sharp criticism of social and economic injustices, not least those resulting from the temporal power of the Church in countries like Italy and France.

There is nothing in these diaries to suggest that they were written by a man who within a few years would be initiated into mysticism. On the contrary, one is struck by the worldliness of Swedenborg's life as a tourist in Paris during the fall of 1736. He went to the theater repeatedly and made careful notes of the names of the actors and dancers: one might ask oneself to what end. On the other hand, he scarcely mentions things that would interest today's reader: the purpose of his long stay in Paris (1736–1738), the people he met, the books he read, etc. Obviously some of this is revealed in other ways, particularly through the tremendous number of excerpts, which were either left in the manuscripts or directly used in the great biological works resulting from the 1736–1739 voyage and the new expedition beginning in 1743; nonetheless, one would gladly have exchanged the names of Parisian histrions for precise information about his learned work.

A journal from the 1743 voyage is still extant, and there is no doubt that it was originally planned to include the same kinds of data as the two earlier ones. However, the notes only relate facts about the route from Stockholm to Groningen and Leeuwarden in Holland via Hamburg before they are broken off abruptly. Thereafter we depart almost completely from the normal world of the senses and are given shocking insights into the *mundus subterraneus* of a dream world. The notes from the years 1743–1744, which were written in a rather crude but expressive Swedish, are usually called the *Journal of Dreams* and constitute one of the most widely known and most frequently quoted Swedenborgian documents; they were discovered and published in 1859 and immediately caused a sensation, even something of a scandal.[9]

Obviously, Swedenborg never intended his diary for publication, but it gives us an exceptional opportunity to see Swedenborg, the man, without his official mask. Although the same is true of the earlier travel journals, this one is especially interesting since it permits us to follow Swedenborg's religious crisis during its most agitated stage. We shall study this document in greater detail later on. An entry dated October 18, 1744, illustrates how the dreamer suffered under the burden of his ambition and self-love and how he interpreted his dreams in the light of his sense of guilt:

> How a big dog, which I thought was fastened, flew at me and bit me in the leg; someone came and held its terrible jaws so that it could do no more mischief. Was because of my having been at the Medical College the day before hearing a lecture and having been rash enough to think that I should be mentioned as one of those who understood anatomy best.[10]

The notes are often cryptic and range between the sublime and the ridiculous, but they bear unequivocal witness to the gravity and intensity of the experience.

The *Journal of Dreams* was written during the years when Swedenborg was struggling to realize his gigantic plans for an exhaustive description of man, which was to be called *Regnum Animale*, or the Animal Kingdom (in the sense of "the realm of the soul," i.e., the human body). The outline and several lists of titles show that the work would have comprised seventeen volumes and that it would carry out the tremendous assignment he had formulated in his short treatise on the problem of the infinite in 1734: *ut ipsis sensibus animae immortalitas demonstretur*, that the immortality of the soul may be proven to the senses themselves, i.e., empirically.[11] The work on this truly heroic task is reflected both directly and indirectly in the *Journal of Dreams*.

One of the reasons Swedenborg's notes on dreams caused such a furor when they appeared in the middle of the nineteenth century was that they contained frank descriptions of sexual fantasies. These were a problem for many high-minded reviewers who went to great length to explain them away. But Swedenborg was not a product of the prudish Victorian era and therefore did not regard sexuality as the foremost work of the devil. It is remarkable, in fact, how little guilt-laden the sexual dreams actually are in a text that otherwise bears the stigma of sin. There may be several explanations for this. We must remember that Swedenborg lived in an age that was characterized by a remarkably frivolous attitude to erotic matters; not even the most hard-working scholar could avoid seeing examples of this in the Paris of the 1730s, and certainly not a frequent visitor to the theater as Swedenborg. This is a general premise, but more specific factors were the anatomical studies of human procreation being carried on at the same time. When a lonely bachelor in his early middle age is working on an investigation of the smallest details of the sexual organs, endeavoring to fit them into his comprehensive vision of the human microcosm, it is scarcely surprising that his speculations continue in more palpable form in his dreams.

The *Journal of Dreams* is evidence of the psychical turbulence of the transitional period. The dreamer found peace in April 1745 when he experienced his illumination, a vision that commanded him to put his scientific work aside and devote himself instead to interpreting the Scriptures. With this transition, we find a new firmness in the descriptions of the visions as compared with the notes on dreams. At the same time, the *Journal of Dreams* clearly prophesied what was to come, not least in the interpretation of the sexual dreams. Swedenborg's explanations of the tempting—or repulsive—female figures are almost Freudian in reverse: he identifies them with various abstract concepts, the sciences, philosophy, etc.

This strange combination of vividly concrete visions and dryly abstract interpretations is characteristic of Swedenborg's theosophical writings. For scholars who are not among Swedenborg's followers, reading his vast exegetic texts can be trying. The vigorous and colorful epic of the Pentateuch and the Apocalypse is translated into a metaphysical and theological code that seems light-years remote from the original. The memorabilia of the spiritual world, which he intersperses among the hermeneutic chapters, appear to the reader as oases in a desert. This interchapter material originates mainly in the notes made from October 1747 to the middle of the 1760s in the so-called spiritual diary, *Diarium Spirituale*.[12] As his exegesis grew more assured and rigid, his description of the visions became more confident and his judgment of persons cited more severe.

This applies not least to people who had been close to him or otherwise influenced his development: to Erik Benzelius who is pictured as the worst of infernal spirits, to Polhem and Charles XII—the latter, one of history's typical bachelors, was doomed to the ludicrous fate of a henpecked husband—and to a series of relatives. The comments on Benzelius are the hardest to understand, even if he, as archbishop of Sweden, represented Lutheran orthodoxy, which the visionary detested. To the modern observer, the bitterness of his reactions seems incomprehensible, even pathological. At the same time, the fact that Swedenborg so precisely and so calmly relegated the ideal and benefactor of his youth to the lowest regions of hell may derive from his profound innocence, his passion for consistency, and the emotional aridity of the intellectual. Swedenborg's hell was intended to illustrate the same belief as Dante's Inferno, that God is love.

Judging by comments of his contemporaries, it was easier to be aware of that belief in personal encounters with the visionary. When in Sweden, Swedenborg lived very simply in his house in the southern part of Stockholm, but he indulged in various exotic plants and other attractions in his garden to amuse children and other visitors. His work on the Scriptures kept him constantly occupied, but his working hours were rather irregular and allowed him to go into society. This he enjoyed, and he always tried to behave like a correct elderly gentleman: that he did not always succeed depended on his being almost the prototype of the absent-minded professor. He conversed in a quiet and erudite fashion, but stammered badly when excited. Since he willingly spoke of his unique spiritual experiences, Swedenborg was in great demand in society, and many people also sought his advice on practical questions which only spirits could answer.

His fame spread all over Europe from the late 1750s; indeed, the cultural climate of the latter half of the eighteenth century was in many respects favorable to mysticism and magic. It is understandable that Swedenborg's image was to a great extent distorted by the early recipients of his message, even though he himself always tried to minimize the significance of his association with spirits and constantly warned others from treading the same path unless they had received a call. He never tried to proselyte; he relied entirely on his many treatises, and often neglected to answer letters of enquiry—occasionally with unfortunate consequences for himself, as in the case of Immanuel Kant, whose derisive and exceptionally summary *Träume eines Geistersehers* (Dreams of a Spirit-Seer, 1766) was in part marked by disappointment at an unrequited approach. Nor were any real Swedenborgian congregations founded during his lifetime, even though he acquired a few disciples who disseminated their personal impressions of him.

It is through his literary production that Swedenborg exerts considerable influence on global culture; any understanding of the man that is acquired outside his works is neither extending nor stimulating. As a writer, he belonged to the most productive in the history of literature, and his intellectual powers remained undiminished until a few months before his death in 1772 at the age of 84. The summary of his theological system, *Vera Christiana Religio* (The True Christian Religion), is perhaps the clearest of his works, and it was written as late as 1771. From the perspective of posterity, the approach to this admirable *Summa* appears to have run a straight and logical course, even though it passes through territory not as easily accessible to others as to the pathfinder himself. The following presentation has no other purpose than to attempt to chart the course of the road followed.

2

Swedenborg's Philosophy of Nature

The year 1745 represents the most distinct line of demarcation in Swedenborg's literary production. It was then that he had the decisive spiritual experience that drew him away from his scientific studies to his work as an interpreter of Holy Writ. But his illumination did not lead to a denial of his earlier contributions or a rejection of their results. On the contrary, it is essential to keep in mind the fact, as emphasized by Emerson and others, that Swedenborg integrated his scientific thinking and experience with his newly acquired religious conviction, and this means that those who want to understand the fundamentals of his system should devote greater attention to his activities as a scientist than to his theosophical period.

From the outset, Swedenborg was interested in a whole series of scientific disciplines, but there is, nevertheless, reason to begin with his speculations concerning physical questions in the broad sense up to and including his *Opera Philosophica et Mineralia* (Philosophical and Mineralogical Works, 1734), which mark both the culmination and the end. We shall then turn to his contributions in the fields of anatomy and physiology. His entire scientific ambition after 1734 was concentrated on grasping the essence of the soul—as already said, the program was formulated so that the immortality of the soul should be proven to the senses themselves—and this meant that the biological researches had their true justification in the psychological area. But it also meant that Swedenborg was soon forced to realize the limitations of language, and it was primarily the struggle with language that led to the illumination. A tripartite disposition thus appears to be best suited to the material, though we must remember that sharp lines cannot be drawn between the areas of research, especially in the case of a man with such a burning and early evidenced desire for synthesis as Emanuel Swedenborg.

Cosmological Problems

England was the natural goal for a young student with an "immoderate desire" for mathematics, as Swedenborg described himself in one of his first letters from that country. This passion could not be satisfied in Uppsala, as was eloquently revealed by the observation made by Petrus Elvius, professor of astronomy, to his former pupil in London in 1711: "What the learned mathematicians think of *principia motuum planetarum Newtonii*, since they appear to be pure abstractions and nothing physical, namely, how one *corpus planetarum* shall exert gravity on another, etc., which appears unreasonable."[13]

But his stay in England did not convert Swedenborg into a disciple of Newton, despite his enthusiastic expectations. He was never able to accept Newton's theory of empty space, nor did he sympathize with the new demands for scientific self-restraint that lay behind the proclamation *hypotheses non fingo*. Newton makes his statement in the *General Scholium* at the end of the *Principia mathematica* when discussing the cause of gravity, which he had not been able to discover from phenomena: "And I frame no hypotheses; for whatever is not deduced from the phenomena is to be called an hypothesis; and hypotheses, whether metaphysical or physical, whether of occult qualities or mechanical, have no place in experimental philosophy."[14] To Swedenborg, the description of a process and the formulation of its law, no matter how exact, were completely inadequate. What he sought in research was precisely what Newton's sober program excluded, namely, a search for the most profound causes of the process. Three decades later, this search was to lead him across the boundary lines of the senses—and of science as well.

The Cartesian philosophy that Swedenborg had adopted in Uppsala was thus not disturbed in any decisive sense during his foreign travels, and essential parts of it remained with him throughout his life. It is not particularly apparent in his early production, however, since he never completed the program formulated in the preface to a paper on the changes in the earth's orbit (1718):

> One should begin with a *Theoria Telluris*, or a treatise on the first origin of the planets, on their separation from their chaos and on their course and resistance in their air: it would also be necessary to compare *Cartesii, Newtonii* and others' views, so that the reader himself could judge which best follows reason and geometry, but since this would be too far-reaching, it requires its own treatise.[15]

After preliminary attempts to define his standpoint in Latin papers at the beginning of the 1720s, Swedenborg presented this treatise in the first volume of his *Philosophical and Mineralogical Works* (1734), called *Principia rerum naturalium* (The Principles of Natural Things). However, it is typical of his general philosophy that, in the tract of 1718, after having apostrophized the pioneers of the new science, he opened his presentation with a survey of the belief in paradise and the golden age expressed in the Bible and in classical poetry and philosophy, according to which there was eternal spring in the beginning of tellurian time:

> Homer (the true ancestor of the poets) made so much of the same time that he likened it to a heaven, placed all gods and goddesses there, made homes for Floram, Pan, Pomonam, assumed that Pallas, Venus, and all other deities of pleasure had their resort and their meetings with mortals there, so that gods and their sons lived together with humans; which means that the earth likened a heaven and the age an era of celestial delight.[16]

One may wonder, whether Swedenborg really knew Homer, since many of the deities carry Latin names instead of the Greek originals; but apart from that reaction, this quotation is obviously an early precursor of the view of history characteristic of his exegesis. In *Arcana Coelestia* (Heavenly Secrets, 1749–1756), the story of Adam in paradise is interpreted as referring to the oldest society in which humans understood intuitively the innermost essence of things and every material object bore witness to spiritual truths and were living symbols of the divine.

It is true that we can find a good many critical statements about Uppsala's dogmatic Cartesians in the works of Swedenborg's youth, but they should not be overestimated. These criticisms express the condescending attitude of the scientist active outside the university toward the academic teachers, an attitude that was common in the days of the Enlightenment and the learned societies. Swedenborg embraced his mentor Polhem's views as they were formulated in a letter in 1716:

> *In summa*, if the learned want to have pleasure and honor from what they teach others they should have better knowledge of much of what is now taught, for Nature in many respects has qualities totally different from what *Cartesius* and almost all his followers mean, which can never be better proved than by a daily experience of mechanics and a penetrating analysis of the causes.[17]

Swedenborg also adopted this combination of empiricism and rational analysis as his research program, but like Polhem he did not stray very far from Descartes. They both adhered to Cartesian views concerning empty space and light, and they also retained his corpuscular model for the structure of matter. When Swedenborg presented his version in *Principia rerum naturalium*, it might seem hopelessly out-of-date almost half a century after *Principia mathematica*, but it should be recalled that Alexander Pope's planned epitaph for Sir Isaac needs to be complemented on one important point:

> Nature and Nature's laws lay hid in night
> God said, Let Newton be! and all was light.

In reality, Newton's theories did not gain any great currency until they were presented in brilliantly simplified form by Voltaire in *Eléments de la philosophie de Newton* (1731); even Newton's light had to be adjusted to suit the eyes of the beholders.

Even though *Principia mathematica* was in Swedenborg's library, he remained faithful to Descartes and took the three main themes of his philosophy of nature from the French rationalist's *Principia philosophiae* (*Principles of Philosophy*): (1) that matter consists of particles that are indefinitely divisible, (2) that these particles are in constant vortical motion, and (3) that the earth and the planets sprang from the solar mass. In the world of material substances, of extension, Descartes distinguished no forces other than mechanical; external form and motion are the only concepts needed to describe it. Matter is entirely uniform and consists of an indefinite number of divisible particles, which group themselves in species forming the three elements: fire, ether, and earth. Empty space does not exist, but the extension of space is governed by the same laws as bodies and thus presupposes the existence of material particles.

Swedenborg admittedly tried to go further than Descartes regarding the divisibility of the particles; but when he was forced to conceive concretely of a limit to this divisibility, he did not arrive at any atomic theory but at a no-man's-land between matter and spirit, namely, at the mathematical point which is an abstract construction. This meant that the series of particles led Swedenborg into the world of mathematics, but not, in his opinion, across the border to the Infinite—for there our worldly geometry and algebra are not applicable. During his scientific period, Swedenborg was concerned with keeping this boundary clear-cut, and he frequently accused others of making impermissible analogical leaps across it, as we shall see later on.

In the Cartesian universe, motion within the material continuum can only be produced through the particles having different shapes and sizes. This thesis can best be illustrated by Descartes's theory of the origin of the earth. He believed that it had originated in the fire of the sun and that the birth process had corresponded to the formation of sunspots, namely, that coarser particles found their way to the surface. In severe cases, a sun could be entirely eclipsed by this foam of particles, whereupon it could no longer remain separate from the surrounding element (the ether), but would be sucked into stronger vortices in the neighborhood. In this way our earth would have been drawn into the enormous vortex of the solar system, in whose rotation it follows. The result is thus a kind of heliocentric theory with reservations, in which the earth actually is at rest and only follows along in the solar vortex.

This brings us to the domain in which Swedenborg made his most renowned contributions as a natural scientist, namely, the planetary theory. Among other things, he modified the Cartesian hypothesis by having the earth and the planets emanate from the mass of the sun itself, from which they had been ejected rather than sucked in toward the center of the sun. He thereby greatly simplified the hypothesis, and his theory can be regarded as a precursor of the Kant–Laplace nebular theory. The question has even been raised whether Swedenborg does not reserve credit for this renowned theory, and it has been suggested that Buffon's *Histoire Naturelle* was the connecting link; there was a copy of Swedenborg's *Principia rerum naturalium* in Buffon's library.[18]

There is no established connection, however, and it may have been a matter either of a relatively obvious modification of the generally known Cartesian hypothesis or of another text read by both men. Swedenborg did not refer in *Principia* either to Descartes or to Newton but rather to classical poetry and thought, including Ovid's repeatedly quoted description of chaos in the first lines of *Metamorphoses.* The words in Genesis about the deep over which the spirit of God moved in the beginning were also related to Ovid's chaos—*rudis indigestaque moles,* formless and confused mass. Furthermore, Swedenborg revived the concept of the great world egg formulated in various cosmogonies. This symbol, which to a man versed in biology offered a unique opportunity to tie together microcosm and macrocosm, inanimate and animate nature, recurs with great poetic effect in the early drama of the creation, *De Cultu et Amore Dei* (The Worship and Love of God, 1745), which marked the end of Swedenborg's career as a natural scientist.

The references to classical sources in *Principia* do not indicate any systematic study of the views of the ancient scholars on cosmological questions, but do point to an important contemporary source that Swedenborg encountered during his stay in England in the 1710s. This was Thomas Burnet's *Telluris Theoria Sacra* (A Sacred Theory of the Earth), first published as early as 1681 but still the subject of heated discussion among the scholars Swedenborg met. Although the work is widely known, it has not been observed by Swedenborg scholars, presumably due to a misunderstanding in Tafel's great collection of documents. In his survey of the manuscripts, Tafel reported that Codex 86 contained quotations on cosmology, etc., from an author called Roumette.[19] But the Latin rubric in the manuscript reads "*ex Bournetto*" (from Burnet), and under it we find the classical and biblical sources quoted in *Principia* (as usual without mentioning the primary source), together with notes on Burnet's method of utilizing the classical myth of the world egg. Swedenborg's only direct reference to Burnet was brief, polemical, and early—1718—and gave no indication of the scope of his influence.[20]

Burnet's influence should probably be regarded as a decisive stimulus for Swedenborg's cosmology, even though he greatly expanded Burnet's perspective. *Telluris Theoria Sacra*, as the title indicates, is a hypothesis of the origin of the earth, while Swedenborg's aim, as already stated, was to explain the origin of the entire planetary system. According to Burnet, God created chaos, which admittedly was a crude, incohesive mass, but nevertheless contained the laws for the development of cosmos; and in it the elementary particles were gradually separated into different layers. The heavier ones sank toward the center, the lighter ones rose to the surface, and the very lightest formed the air. Together with dust particles in the air, the oily outer layer of the water particles in chaos formed the crust of the earth that at first was smooth and uniform. The later changes in the surface of the earth, according to Burnet, were caused by the Deluge, which shattered the crust. In general, Swedenborg agreed with Burnet, even though he had a different opinion of the actual technical workings of the Flood. Finally, Burnet made an elaborate comparison between the newly created earth and the world egg of classical mythology. It should be added that, while he restricted his theory to the earth, Burnet also declared that he found it probable that all the planets of our firmament—*ejusdem solis alumni,* fosterlings of the same sun—sprang from the same chaos.

This is precisely what occurred according to Swedenborg's planetary theory, and he thus made the planets the children of the sun, which unquestionably is close to Burnet's view. Now, Buffon was a great admirer of Burnet and analyzed his theory in *Histoire Naturelle,* which may explain the similarities between the Kant–Laplace nebular hypothesis and Swedenborg's theory. The many references to classical speculations in Swedenborg's cosmological contexts caused certain scholars to regard direct impulses from antiquity as decisive. An example of this is the assumption by Alfred Stroh, the respected American Swedenborgian scholar, that Ovid's description of chaos was his principal source.[21] However, a comparison between *Telluris Theoria Sacra* and *Principia rerum naturalium* clearly shows that Burnet's work holds that position.

This applies both generally and in detail, and it is not unlikely that an even more profound influence than the purely cosmological should be recognized. Coleridge once planned to transform Burnet's work into an epic poem, and he refers to it in his *Biographia Literaria* as proof that the noblest kinds of poetry need not have the outer form of a poem.[22] This poetic charm, noted by Coleridge, probably resulted from the author's view of poetry and myths as media of an esoteric wisdom, in this case from an age before both fables and written history, to use the somewhat schematic differentiation of the day. In any case, one can state hypothetically that Swedenborg's warm appreciation of the cognitive value of the myths of the golden age and classical poetry is related to his reading of Burnet as a young man, and it is noteworthy that he himself experimented with a poetic "mythical" form when he tried to summarize his knowledge in the natural sciences in the unfortunately—but maybe inevitably—unfinished *De Cultu et Amore Dei.*

Let us pause before his vision in that book of what happened after the sun had shattered the shell of the great world egg:

> On the bursting of this immense repository there sprang forth large masses, equal in number to the planets visible in this universe, and resembling our earth, but which being yet without form, and not balanced in any ether, pressed upon the great surface of their parent; for no force was as yet operative to carry them in another direction. Thus they lay scattered like suckling masses near the burning bosom of their father, and, as it were, at his teats. But presently when the sun, the folding-doors being unlocked and the gates thrown open to the empty universe, had begun to cast forth fiery exhalations from his now full and swelling mouth, and to distend it with his powers and forces, he first filled the neighboring and presently the more remote distances with auras and thus with spaces. Hence arose ether, which being diffused around the sun, and at the same time also around the masses which encompass him, wrapped the latter, as it were, in swathings or spires, and encompassed them with spheres suited to the mobility of each. In the circumferences of their spheres he placed a pole, which he drew into perpetual orbs, and from them produced a central gyration, in which the mass was involved. Hence it came to pass that those bodies, being as yet fluids, and as it were molten, assumed an orbicular form from the concourse of so many centripetal forces. These now became orbs, and of no weight, as it were, because in centres, and being conveyed and put in rotation by the surrounding ether, they first began to creep and then to walk around the sun, and presently, like little children, to dance and leap, and by quick and short circuits to make a commencement of years, and a rotation of days, and thus to enter upon their periods.[23]

The Mathematical Point, Particles, and Elements

Even though Swedenborg never gave the same grand framework to his speculations on the beginning of the particle series as to its completion, when Earth and the other planets were born, they called for at least as great imaginative powers. As early as in his first small essays on what would now be called theoretical physics, in *Prodromus principiorum rerum naturalium* (A Specimen of a Work on the Principles of Natural Philosophy, 1721) and in *Miscellanea observata circa res naturales* (Miscellaneous Observations on Natural Things, 1722), which were preliminary studies for the magnum opus of 1734, *Principia*, Swedenborg allowed the series of particles to originate in the mathematical point, sometimes called *punctum naturale*, the natural point, sometimes *punctum Zenonis* after the Eleatic philosopher. All geometric figures could be restored to that point, if only motion existed, and the Creator himself introduced motion into the universe.

At the outset, this point obviously only served as an auxiliary concept, and Swedenborg soon left it behind him. However, he dwelt on it at greater length in the *Principia*. The Creator had introduced in the point a *conatus* to motion—the scholastic notion can be rendered as "striving," and Swedenborg probably borrowed it from the great university philosopher of the day, Christian Wolff, a disciple of Leibniz, with whose metaphysics he had become acquainted just before the final editing of *Principia* in 1733—but it could not be described in geometrical terms. This lack of mathematical proof sorely tried Swedenborg, and he expressed a hope that others would find a more adequate means of making it intelligible. Until someone did, he had to be content with images and symbols that at times gave admirable results. Characteristically, he once more returned to classical mythology. This time it is the Roman god Janus with his two faces who illustrates the ability of the mathematical point to look at two worlds simultaneously, at the infinite and at the physical universe. Janus was the god of the archway and of the door, and this is the function from which Swedenborg proceeded when he described the purpose of the mathematical point in the following passage:

> By this point, as by a door, we are introduced into the world; and admitted into a kind of geometrical field, where ample scope is afforded for the exercise of the human understanding. As soon as, through the medium of this point, an entrance is found, into the finited universe or the world, man instantly begins to have a knowledge of himself, to perceive that he is something, that he is finited,

> mechanical, nay, even a machine: in other words, by this point we are introduced into the world, and into its law, that is, into geometry, which could have no existence prior to the existence of the point. Nature itself also commences with the same point: to this it is indebted for its birth, that is, for its conception and exclusion as from the womb; and from this it first receives what may be called its life, and its forces under their several modifications. Wherefore the world is incipient at this point, and with the world nature itself; or, what amounts to the same, nature has its incipience by it, and the world with nature. On these grounds and considerations, our point may be said to be the medium between what is infinite and what is finite.[24]

This visually understandable description, which in its ability to give concrete life to an utter abstraction—and in its verbosity as well—is so typical of Swedenborg, is still grounded in the thought of a natural scientist of the Cartesian denomination. That which existed at the other exit of Janus' archway, the infinite, still lacked definition, but the methodological discussion in the introductory chapter of *Principia* reveals how a melancholy Swedenborg imagined that the first human beings would have intuitively experienced the infinite. Nature in itself is nothing, did not create itself, but is only a product of the infinite via the point and motion. In the age of innocence, when there was complete harmony between the physical senses and the soul, human beings were constantly aware that nature is but an image of eternity; but this harmony was destroyed by the fall of man, channels of knowledge were broken, and the order of the universe was disrupted. Now we could only guess at the truth, and our obscure thoughts could only be expressed in analogies and images.

At the same time, Swedenborg had no reservations about devising an extremely detailed series of hypothetical particles and elements between the mathematical point and the material bodies about which we learn from our senses. However, in this entire vast exposition only two principles are active in constant repetition and variation, namely, the point and the force introduced into it. As behooved a disciple of Descartes, he interpreted this force as the most perfect form of motion, the spiral. Together with the immaterial point, motion created the first simple substance, which in Swedenborg's Latin was called *simplex finitum*, the simple finite. It was passive in itself, nature's cornerstone, but it had the ability to emit a free force called *activum primi.* From this force was created the second substance, which formed the first element of nature. It in turn released a free force, *activum secundi,* and together these forces formed the solar ocean. The next stage was the third substance and the second element, the magnetic. And so the process continued up to the terrestrial bodies and the organisms, or possibly even further.

It is thus the natural point that is multiplied to ever greater powers in this all-inclusive series. The elements were assumed to be made up of a kind of bubbles *(bullula)*, whose surface consisted of the next higher finite, while the interior was filled with the preceding *activa* together with the *activum* of the stage itself. The surface of the magnetic element was thus formed by the third finite and its interior by *activa primi et secundi.* In this way, each *bullulum* became a reflection of the entire universe. Even the most concentrated description of this grandiose system must convey the impression of a great imaginative gift, especially sensitive to uniform models, analogies, correspondences. Obviously a vision of such compelling force includes a tendency to cross boundaries, to appropriate increasingly vast domains. In *Principia,* however, Swedenborg merely gave a hint that the series might have a continuation. Perhaps higher orders of *activa* exist that our senses cannot apprehend. What he is referring to must surely be angels and spirits, in whose existence he shared the vague belief of his contemporaries. Not many years were to pass before these spiritual substances became the visionary's subjects of research and his daily companions.

The Infinite

In the same year that *Principia* was published, Swedenborg presented a small treatise on the notion of the infinite and its relations to physical reality. He was aware that this might be considered an investigation beyond his competence, since the Scriptures had already provided the answer—we should remember that Swedenborg was brought up in an environment where the theologians strictly guarded their domains and had the right to dismiss unceremoniously all Cartesian intruders—but he said he wanted to reach a solution that would be acceptable even to modern sceptics and might prevent grave error. He presents a clearly disposed and stringently executed argumentation, which begins by discussing evidence that the infinite actually exists. Then, after acknowledging the physico-theological proof of the existence of God, he gives a critical survey of a series of different beliefs in God and philosophical attempts to define the notion of the infinite. The principal objection here is that, by endeavoring to treat the infinite with the usual methods of finite thought, one lands in pantheism or in nature worship. The one who best survived this philosophers' ordeal was Aristotle. This may seem strange in a thinker who otherwise adhered so closely to Descartes, the arch-enemy of Aristotelianism, but it illustrates a change in the intellectual climate, which there will be reason to discuss later.

Finally, all philosophical arguments proved inadequate and Swedenborg was compelled to take refuge in Revelation. The ardently desired link *(nexus)* between the infinite and the finite was the Son of God who, together with the Father and yet separate from him, created the universe. No reference is made to any particular part of the Bible, but he must have had in mind the preamble to the Gospel of John, the most philosophical of the Gospels; the divine *Logos* is hinted at as the first link in the series of nature. For this alone, *De Infinito* represents one of Swedenborg's most significant approaches to the spiritual world during the first part of his life.

Before he could proceed, however, vast research problems remained to be solved. Significantly enough, the tract on the infinite was published together with a paper on the interaction between soul and body. After 1734, Swedenborg turned against organic nature, in the last resort towards man. This meant a shift from the origin of the series of particles to its final point and objective, from the birth of the planetary system to that of the organisms, from oscillating light and sound to vibrating nerve fibers—a tremendous widening of perspectives in which the already structured model nevertheless remained intact.

3

From Inorganic to Organic Nature

The Seventeenth-Century Background

In the Cartesian philosophy of nature, the world of matter is separated from the spiritual by insurmountable logical barriers. Descartes included in matter the biological organisms, all of which possess the attribute of extension. It is therefore consistent that Cartesian biology came to be marked by a mechanistic point of view and that it treated the organisms as natural phenomena, which did not differ in principle from inanimate objects; and it is equally logical that the Cartesian presentations of the philosophy of nature were generally concluded with biological and anatomical passages. The master himself in *Traité de l'homme* (Treatise on Man) applied his mechanistic physics to human physiology and to a certain degree to human psychology also.

As a basis, Descartes used the results of the anatomical research initiated through Andreas Vesalius' *De fabrica corporis humani* (On the Structure of the Human Body), published the same year as Copernicus' macroscopic masterpiece *De revolutionibus orbium coelestium* (On the Revolutions of the Heavenly Spheres), 1543, and which, in Descartes's lifetime, had as brilliant a representative as William Harvey. When Harvey unassumingly published the renowned *Exercitatio de motu cordis* (Anatomical Dissertation upon the Movement of the Heart, 1628), his studies on the circulation of the blood, he torpedoed the Aristotelian belief that various kinds of "vital forces" determine the life of the organism by proving that the motion of the blood has purely mechanical causes, that the heart simply functions as a pump.

The growth, triumph, decline, and fall of mechanistic biology can be traced over the period 1670–1745.[25] Descartes's system began to gain general acceptance around 1670, and by the turn of the century, mechanistic biology had become irrefutable dogma for most French scholars. In this area, as in many others, France set the pace for the learned world of Europe. This view of the structure of the organisms obviously favored anatomy as a science. The mechanism of life could be determined only with the help of the concepts of motion and form. Hence, it was essential to determine the pattern of motion and geometrical figuration of the organism; the anatomists did their best to detect "machines" in the human body resembling those that could be made from inanimate matter.

This mechanistic dogma was one important stimulation for the development of biology, the endeavor to build on observation and experimentation being another one. Microbiology became increasingly significant in the second half of the seventeenth century as a result of the work of a number of microscopists headed by the self-taught Dutchman Leeuwenhoek. With the help of ground lenses, fantastic views of a hitherto unknown world were opened, and it is understandable that many were intoxicated by the sight of the incalculable riches of nature and tried to find analogies and correspondences everywhere. During the latter part of the period, however, this empiric ambition began to lead to new theories replacing the mechanistic dogmas, at least the purely Cartesian. The principal stumbling block of these theses was always the same: the explanation of the origin of new organisms. The laws of motion and the geometrical patterns introduced into matter proved untenable as explanations of the genesis of highly complicated organisms.

Three main theories concerning the origin of organisms were dominant during the years 1670–1745. The first was "ovism," the belief that all living beings are born from eggs through the contributions of two parental individuals: its slogan, *omne vivum ex ovo*, everything living (comes) from an egg, was called by Linné in 1739 "the first article of faith in natural history." Leeuwenhoek was the father of the second theory. In 1677, he discovered spermatozoa, which he looked upon as a kind of tiny animals, *animalcula*. To begin with, however, he did not grasp their significance in regard to fertilization, but confined himself to attacking the current belief on one point, to wit, the belief that the fetus is not developed in the egg before it leaves the ovary and that the sperm is merely a transmitter of the fertilizing spirit *(aura seminalis)*. Instead, he believed that the fetus is pre-formed in the thicker parts of the seminal fluid and thereby implanted in the uterus. This brought him into conflict with the ovist theory; and, in heated debate with its proponents including members of the Royal Society in London, he presented in 1683 an animalculist theory, according to which the spermatozoa are pre-formed fetuses that attach themselves to the uterus. As soon as a single one of the great mass has found a hold, which can only be done in two or three places, it repulses all the others and immediately begins its long process of transformation.

Leeuwenhoek's theory was disseminated by the learned journals, which at that time began to play an important role as media for the discoveries of the new sciences, such as Le Clerc's *Bibliothèque universelle et historique*, *Journal des Savants*, and the rest; but it was severely criticized on several points. Many doubted that these *animalcula* existed. The idea that the fetus was pre-formed in the egg was a hurdle, and it was difficult to grasp that such masses of spermatozoa were required when only one would survive. This objection was of theological significance, since the theory seemed to imply that God would unnecessarily create an infinite number of diminutive beings who would never achieve independent life. Leeuwenhoek himself believed that the spermatozoa were formed in the testicles and that they constituted a special genus of animal—in French, his theory was called the vermist or worm theory—and that they thus were not infinitely tiny beings but only bore such within themselves. This explanation obviously did not solve the problem, and Leeuwenhoek's specific animalculist theory was more or less abandoned with his death in 1723.

At about the same time, the interests of the anatomists were diverted in other directions. However, the third generation theory, which developed from their discoveries, survived—namely, the notion that the seeds and eggs contain infinitely tiny individuals who merely grow in size during the fetal period. This theory dated back to another great Dutch scholar, Jan Swammerdam (1637–1680), who formulated it in his work *Historia insectorum generalis* (General History of Insects, 1669) and, with his religiously mystical temperament, drew several important theological conclusions from it. The pre-existent human seed was present in the egg, and all human eggs had existed in the ovaries of Eve, the mother of humankind. This also meant that the human race would die out once the finite number of eggs had been exhausted, and that Eve's fall affected all her progeny in a biologically concrete sense, which provides an extremely tangible explanation of the concept of original sin.

The theory was based on microscopic observations, including those made by Marcello Malpighi (1628–1694), the scientist so frequently quoted by Swedenborg. It spread rapidly and was widely accepted through the philosopher Malebranche, who included it as early as 1674 in his famous work *Recherche de la vérité* (Search for Truth), which also strongly influenced Swedenborg; the pursuit for truth was started by the famous Oratorian father from Cartesian premises, which was not least important to Descartes's disciple Swedenborg. The preformation theory exactly suited Malebranche in the sense that it deprived nature of all independent activity. Only such things developed in nature as were implanted in it by the Creator before the beginning of time and in utter subservience to the immutable law of order. It further meant that the mechanical principles present in the world did not explain its creation. The world, inanimate as well as animate, could not have come into being from mechanical causes but presupposed a Creator, who admittedly allowed all to take place according to the laws of mechanics and geometry but who gave this conformity to law a fixed ultimate objective.[26] The same teleological view was held by Leibniz

(1646–1716), who also took up the pre-existence theo[...] anatomists; but through his special metaphysics, the pref[...] concept eventually was replaced by a dynamistic approach, [...] Swedenborg enters the picture in an interesting way.

Swedenborg's Biological Speculations until 1734

It is natural that a versatile spirit with Swedenborg's Cartesian training would try at an early stage to penetrate the life sciences. Swedenborg's first attempts were to explain the nature of human sensory perceptions. The initial example was a short paper in the last issue of the periodical *Daedalus Hyperboreus* (1718), entitled *Bewis at wårt lefwande wesende består merendels i små Darringar thet är Tremulationer* (Proof that Our Vital Essence Consists Mostly of Small Quiverings, i.e., Tremors). Here Swedenborg extended his observations on the waves in different media to apply to the human body, and we see how heavily he leaned on Polhem's studies of the nature of sound, which were published in earlier issues of the periodical.

All sensory reports, both external and internal, consist of tremors of the nerve fibers that pass on to what Swedenborg rather vaguely calls the brain membrane, and from it such "quiverings" in turn spread to the entire body. He also believed that they could reach beyond their own organism. He thus assumed the same effect as a vibrating string on a musical instrument can have on the other strings tuned to the same key:

> It often happens that a person joins another's thought, so that he gets a presentiment of what the other does and thinks, i.e., that his membrane quivers because of the other's quivering, as one string because of another's, because they are tuned to the same pitch.[27]

In these terms, the organism most closely resembles a complicated musical instrument with strings, membranes, and soundboard, which should, of course, be borne in mind when trying to evaluate the recurrent use of the term "harmony" by Swedenborg; the word may often have a musical-technical import in addition to other meanings. The *Daedalus* essay was in a sense a first draft, and Swedenborg soon developed it into a more ambitious scientific effort, which in 1719 he submitted to the Collegium Medicum in Stockholm for evaluation; only an incomplete copy is extant.

Swedenborg himself declared that this work stemmed from the Italian anatomists Baglivi, Borellus, and Cortesius. The last name has hitherto been misread as "Cartesius," which is easily understood because of the similarity and because the point of departure of the study was a Cartesian theory of motion as the cause of life. Swedenborg considers tremor to be the most subtle of all the forms of motion in nature. There are various degrees of tremor, however, from the coarsest and most distinct as in the lungs, heart, and—characteristic of the anatomy of the day—the brain. Swedenborg distinguishes between three types of tremor: undulation, true tremor, and *contremiscens*, or sensation.

True tremor embraces the range of vibration that gives rise to sound effects, which Swedenborg fixes at approximately 30 to 200 vibrations per second; as we know, modern measuring methods set the upper limit much higher, approximately 20,000 vibrations per second. The sensations, in his view, are movements of the same type but with such high vibratory levels that they cannot be measured or heard. The essential point, of course, is not the measurability but the fact that Swedenborg was endeavoring to interpret all sensory functions on the basis of a single mechanical model.

The propositions were proved in a manner characteristic of Swedenborg's entire methodology. He himself described this as "building on an infinite number of experiments, taking advantage of the labor and expenditures of others; namely, working with one's brain on what others have worked with their hands."[28] With the help of the great anatomists of the seventeenth century, particularly Raymond Vieussens and Thomas Willis, he first described the nervous system from the medulla oblongata and the medulla spinalis to the furthermost extremities and then the various membranes of the body.

The latter, he believed, consist of nerve tissues and form cuticles that enclose all the organs, but, most important, all the vascular systems, the lymphatic vessels as well as those of the blood; and, to the extent that the vessels are filled with fluid, the cuticles are kept distended. The latter is essential if they are to register and transmit tremors; but, in addition, the membranes are related in various ways to the skeleton, which is necessary if the tremor is to be sufficiently powerful. This is particularly applicable to the most advanced and most important membranes, the dura mater and pia mater, the hard and the soft cerebral membranes. Here Swedenborg is uninhibited in his use of the analogy with musical instruments. The skull, he claims, is porous for the same reasons that porous woods, like cedar and pine, are better suited for use in string instruments than harder woods, in order more easily to give resonance to the vibrations of the cerebral membranes.

Swedenborg was particularly interested in the lymphatic system, the origin of which he considered to be in the brain. In the same way as the heart pumps blood, he believed that lymph flows from the brain and circulates via the nerves to the various membranes and from them back to the brain. The vibrations can thus emanate from this flow if it is a question of the brain's giving orders to the body, or else be transmitted to it from the membranes around the lymph vessels in the case of external influences on the body. This notion is, of course, closely related to the Cartesian hypothesis of the "spirits of life," *spiritus animales*, as mediating organs between body and soul. Purely locally, Descartes presumed that the exchange of reports and orders takes place in the pineal gland (*glandula pinealis*), which was one of the sensational discoveries of the new brain anatomy. In his essay of 1719, however, Swedenborg was patently unwilling to accept the conventional view of these spirits of life and worked energetically to get away from what he regarded as occult assumptions and to arrive at a thoroughly mechanistic model.

This becomes even clearer when we come to Swedenborg's next attempt to create a new psychophysical theory. It appeared in 1734 and is based on the expanded propositions of the series of elementary particles presented in *Principia rerum naturalium* the same year. An essay on the mechanism of the activities of soul and body was published together with the study *De Infinito,* mentioned above. It proceeded from the daring premise that the soul is a finite entity and thus subject to mechanical laws like the body. This does not mean that the soul can be identified with any of the subtle particles described in *Principia*, but in the draft manuscript, Swedenborg draws a parallel between the series of particles and the organism's membranes at lower levels. Not even the purest and simplest of these particles is subtle enough, however, to serve as building stones for the soul.

Despite this reservation Swedenborg was naturally aware that his mechanistic view of the soul might lead to misunderstandings and considerable difficulties, for example, in explaining the highest powers of the soul and ensuring its immortality. He found, however, that it was even more dangerous to content himself with man's ignorance of the nature of the soul and to resort to such solutions as offered by the "spirits of life." That he regarded this as an empty phrase appears from ironic statements about earlier scholars's efforts to bridge the abyss between soul and body created by Descartes:

> It was on this account that they called these emissaries, "spirits," adding also the predicate, "animal," so as to have some words to express the medium, and at the same time to convey that these spirits were as thoroughly unknown as the spirit of man apart from the body. They selected the expression because they could not deny the nexus, and yet were in the dark as to its nature.[29]

Why not try, instead, to apply the mechanistic laws to this area also and in that way allow psychology to make the same progress achieved by geography and anatomy in recent times?

The solution to that problem required a gigantic research program, which even today remains far from complete, if even begun, and Swedenborg was largely aware of this. His essay has, indeed, the character of a test paper in a scientific seminar, a draft of a working hypothesis. Swedenborg appeared as the proponent of a very audacious and modern scientific ideal, and there is no doubt that he saw himself as a pioneer. Had Swedenborg's writing been interrupted at this point, we would probably have regarded his project as an extreme example of the worship of reason of the Enlightenment and of scientific arrogance, possibly also as a precursor of such materialistic systems as the radical French Enlightenment produced in La Mettrie's *L'homme machine* (The Man Machine, 1748). Diderot's *Conversations between d'Alembert and Diderot* and *D'Alembert's Dream* often bring to mind Swedenborg's quivering nerve fibers, which is understandable, of course, in view of the common Cartesian background.

Even though Swedenborg's intent was the direct reverse of the materialists, the associations with them are justified, since their consistently sensualist method corresponds to an important degree with Swedenborg's own ideal at the time. As he then experienced the problem, the chief risk of using unproven "spirits of life" was that modern scientists would be tempted to deny the independent existence of the spiritual. To avoid such disasters, Swedenborg set up the incredible objective for his own research that we have met twice before: *ut ipsis sensibus demonstretur animae immortalitas*, so that the immortality of the soul may be proven to the senses themselves.

The outline of this enormous research program was presented in the form of thirteen propositions. The first seven give general epistemological premises, repeating, among other things, the general theory of elements from *Principia*, but with the eighth Swedenborg proceeds directly to the human body. He begins with the construction of the sense of hearing, referring clearly back to the tremor essays of fifteen years earlier. Every part of the body is assumed to be covered by a membrane. This is a repetition of the proposition of his studies as a young man, but it must be based in part on analogies, since not even the strongest microscope could yet reveal the most delicate membranes.

After emphasizing the part played by the bodily fluids in distending these membranes, Swedenborg sets forth his tenth proposition: that exact harmony must exist between the membranes and the elements if the elements's movements are to be transmitted to all parts of the organism. Here he states outright, although the details were left for a later investigation, that "the vibratory or undulating motions in the elements enclosed in the frame, are in the truest sense those animal spirits that are said to obey the volitions of the soul, and that realize whatever is desired by the body and the soul conjointly."[30]

Since the elements thus play the leading roles in this psychophysical drama, the laws governing their movements also determine the reactions of the membranes. As established in the eleventh proposition, these laws are characterized by harmonious proportions, and the movements are transmitted quickly and surely to the extent that the bodily membranes are in harmony. If this is not the case, the soul can only be reached by obscure and painful reports, to which it can become habituated; as a result, the body's influence on the soul will become increasingly strong. This is extremely dangerous, the reason being that, when a soul in this state is finally released from its body, it can only experience discomfort and pain in contact with an otherworld existence of incomprehensible perfection and harmony.

This warning is a most interesting harbinger of Swedenborg's theosophical period. The distribution of spirits between heaven and hell in the spiritual world, to which souls come after the death of the body, is determined by the choice of the spirits themselves: an impure spirit cannot exist in the realms of the blessed, but must voluntarily seek out its sullied peers.

The question then arises as to the extent to which an individual can influence his own membranous system. The thirteenth and last proposition treats the problem of the soul's site in the body. Swedenborg explained that it is not to be found in any special gland or single membrane but that it must be in the organ in which the membrane has achieved the highest level, in other words, in the brain, primarily the cortical substance but also parts of the marrow such as the medulla oblongata. Such a relationship to the brain would inevitably mean that souls are not identical, but vary according to the physiological foundation. It should be mentioned, however, that Swedenborg, in a short sketch of the same period entitled *Generaliter de motu elementorum* (Generalities on the Motion of the Elements) claimed that the shape of the fibers and the muscles of the body and the cerebellum are predetermined, while the rest of the brain's structure is formed by upbringing and the use to which it is put.

Experiences thus literally make their marks on the brain substance. Against that background, we can sense the earnestness with which Swedenborg was searching for the truth about the soul: the knowledge he could acquire would finally determine the course of his eternal life. The extremely detailed discussions and the tremendous accumulation of facts are comprehensible in the light of his need to create a line of reasoning based on as solid associations as the elements or the membranes of the body—otherwise the brain would be structured in a way that would allow the soul to cultivate dangerously unharmonious errors.

The short essay obviously had a very concrete intellectual foundation, even though Swedenborg refrained from presenting especially precise details. Meantime, several of his manuscripts go much further; but it is difficult to decide how convinced he was of the truth of his speculations—like other scholars, he should be allowed the right to be judged according to the works he felt he could publish. Nevertheless, they are of great value as illustrations of the force of his purely natural philosophical inspiration.

One sketch experiments with the construction of the soul—called here *anima rationalis*, an Aristotelian term that later became Swedenborg's usual designation of the highest spiritual function—proceeding from the actives of the first and second finites that, according to the *Principia* model, form small spaces on whose surfaces the passive elements are collected. These spaces and surfaces together form an extraordinarily subtle membrane, which he thus believed enclosed the soul. Animals possess the same membrane, according to the sketch, but their "soul space" is made of grosser material, one of the ingredients of which is the magnetic element from *Principia*. This explains their instinctive recognition of north and south, their miraculous sense of direction, an opinion that Swedenborg maintained in published works also.

In humans, the soul can neither be destroyed nor can it disintegrate, even though as a finite being it possesses extension; but it can be compressed into a sort of ball, which occupies the smallest imaginable space, and it can easily expand again to its original form or to some other shape. When the body dies and turns into dust, these soul balls are removed by angels. Swedenborg does not specify, however, whether they later regain their former appearance in a particular locality, but this would be quite in line with his other reasoning, at the same time as it is an early augury of Swedenborg's notions of the spiritual world after 1745 and the "substantial" bodies of the spirits.[31]

In another manuscript, consisting of annotated excerpts from the philosopher Christian Wolff's newly published *Psychologia empirica* (Empirical Psychology, 1732), Swedenborg went furthest in concretion, so far, in fact, that he even drew a sketch of the membranes of the soul. He assumed that the innermost parts of the soul are spiral in form and can thus be regarded as a sort of rigid Cartesian vortex. The reason for this shape is that it permits different influences, depending on the nature of the tremor that reaches it. This means that there are infinite possibilities of individual variations, in the same way as there is no such thing as two exactly identical musical instruments, even though their mutual harmony is perfect—still another example of Swedenborg's tendency to describe the activity of the soul in musical analogies.

Above the spiral convolutions of the soul, there is another excessively sensitive membrane, said to enclose the first element of nature, created by the first force according to *Principia*, and then still another containing the second magnetic element. On top of that there is a third membrane consisting of the third element of nature that, according to Swedenborg's reinterpretation of Descartes's physics, constitutes ether, and after that follow two more membranes filled with subtle fluids. The whole is assumed to be enclosed in a membrane composed of blood vessels. Finally, Swedenborg asserted that future research would confirm his hypothesis, and his train of thought is well illustrated in the following passage: "If we had the microscope, we might be able to see the entire structure both of the soul and of the spirit"[32]—a microscope of sufficient strength and clarity, that is! Like so many of his contemporaries, not least his own models and mentors, Swedenborg was fascinated by the possibilities offered by refinements in microscopy. But unlike others, he did not have the patience to wait for stronger lenses. Taking advantage of poetic license in *De Cultu et Amore Dei* (Worship and Love of God, 1745), he let an angel use its own body to show an astonished Eve how the nerves of the body have their origin in the cortex of the brain; we shall return to the visions of the celestial microscopist in the proper context.

Works of the Years 1734–1745

The works published in 1734 mark both an end and a beginning. The theory of elements and particles had reached its culmination, and Swedenborg's main endeavor now was to apply it to the realm of the living. At the same time, he outlined a gigantic research program with the aim of achieving scientifically proven knowledge of the soul according to mechanistic principles. Swedenborg immediately began to work on this program while he was still traveling in Germany in 1733–1734. The same volume of manuscripts, which contains notes on his travels and the annotated Wolff excerpts recently mentioned, includes several hundred pages of excerpts from biologists and anatomists: Vieussens, Winslow, Ridley, Verheyen, Heister, and many others. Together with quotations from great scholars such as Swammerdam, Boerhaave, Leeuwenhoek, and Malpighi, among others, these notes constitute the bio-scientific foundation for the physiological and psychological works of the 1740s, up to and including their poetic culmination *De Cultu et Amore Dei.*

Swedenborg had not the patience to remain long in his Swedish isolation, particularly since the inheritance from his father, who died in 1735, offered him still better financial resources for travel; and in May 1736, he requested three years's leave of absence from his position to allow him to continue his researches, which, as he wrote in his application to the king, required access to foreign libraries and conferences with the scholars of Europe. He began his voyage in July 1736, traveling first to Paris where he spent a little less than a year and a half. He then moved on to Italy and from there to Holland, where he completed and published two great volumes in a planned series originally to be entitled "The Animal Kingdom, both Physical and Psychological, or a System of Natural Principles and Phenomena concerning the Two Parts of Man, that is, concerning the Soul and the Body, and concerning the Causes and Effects, the Actions and the Passions of each; and, at the same time, concerning the Elementary World wherein they live; Explored and Demonstrated both from Principles and from Experiments Physically, Chemically, Anatomically, Mechanically, and Philosophically."[33]

In reality, the first works carried a different title, *Oeconomia Regni Animalis* (*The Economy of the Animal Kingdom* (that is, the organization of the soul's kingdom), with the subtitle "considered anatomically, physically, and philosophically,"and they were intended to be a part of a series that would elucidate the human microcosm as completely as the original title suggested. Judging by a dispositional outline in one manuscript, it would lead up to a theological perspective or, more precisely, to the purpose of the creation so self-evident to a scholar with Swedenborg's background, namely, the kingdom of God.[34] But it is not hard to grasp the enormous work implied by the title, and it is equally clear that the plans were impracticable.

The period up to 1745 was filled with hectic work on at least three different projected publications. Besides the *Oeconomia* series already embarked on, Swedenborg had a plan for a number of short treatises, which would take up one problem at a time, and a great project to comprise not less than seventeen volumes, which would be given the title proposed for the original plan—*Regnum Animale*, "The Animal Kingdom," in the sense of the human body—and which would present the entire physiological and psychological structure of man based on detailed investigations of each separate organ. Swedenborg had time to publish only three volumes of this series before he was diverted from this work and set out to summarize his philosophy in poetic form in *De Cultu et Amore Dei* (The Worship and Love of God). This work, too, remained unfinished. In April 1745, the author received the decisive call to devote the rest of his days to interpreting the Bible according to such principles as prevailed in the golden age of paradise.[35]

Even a condensed summary reveals that the period between 1734 and 1745 was one of intense research and constant shifts in perspective. It will be necessary to confine ourselves to what seems to be particularly essential, and this means that we must examine how Swedenborg carried out his psycho-physical speculations by profound studies of the anatomy and physiology of the brain. That we are justified in doing this appears from a quick look ahead at a theosophical work from 1763, *Sapientia Angelica de Divino Amore et Divina Sapientia* (Divine Love and Wisdom). The book ends with the following statement on the human fetus, which is based on what the angels had told the author:

> The initiament or primitive of man as it is in the womb after conception, no man can know, because it cannot be seen; and also it is of spiritual substance which does not fall into vision through natural light. Now because there are some in the world whose bent it is to explore even the primitive of man, which is the seed from the father, by which conception is effected; and because many of these persons have fallen into the error of supposing that man is in his fulness by his first conception, which is but his beginning, and that afterwards he is perfected by mere enlargement of growth; therefore it has been discovered to me what that inchoate or first thing is in its form. This was discovered to me by the angels, to whom it was revealed by the Lord. They, because they had made it a subject of their wisdom, and it is the joy of their wisdom to communicate to others what they know, by permission presented the initial form of man in a type before my eyes in the light of heaven. Which was as follows.
>
> There was seen as it were a least image of a brain with a subtle delineation of somewhat of a face in front, with no appendage. This primitive in the upper gibbous part was a compages of contiguous globules or spherules, and every spherule was compacted of others still more minute, and every one of these in like manner of spherules most minute. Thus it was of three degrees. In front, in the flat part, a kind of delineation for a face was apparent. The gibbous part was covered round with a most fine membrane or meninx, which was transparent. This gibbous part, which was a type of the brain on the least scale, was also divided into two as it were beds, as the brain on

> the greatest scale is divided into two hemispheres; and it was told me that the right bed was the receptacle of love, and the left bed the receptacle of wisdom; and that by wondrous couplings they were as it were consorts and intimates. Moreover it was shown in the light of heaven, which shone with favouring effulgence, that the compages of this little brain within, as to make and fluxion, was in the order and in the form of heaven, and that its outer compages was in direct opposition to that order and that form.
>
> After these things were seen and pointed out, the angels said, that the two internal degrees, which were in the order and in the form of heaven, were the receptacles of love and wisdom from the Lord; and that the exterior degree, which was in direct opposition to the order and the form of heaven, was the receptacle of hellish love and insanity; because man by hereditary corruption is born into evils of every kind; and that these evils reside there in the extremes; and that this corruption is not removed unless the higher degrees be opened, which, as was said, are the receptacles of love and wisdom from the Lord. And because love and wisdom is very man, for love and wisdom in its essence is the Lord, and this primitive of man is a receptacle, it follows that thence in that primitive there is a continual travail into the human form, which also it puts on successively.[36]

This is one in an endless row of examples of how Swedenborg's researches on the physiology of the brain gave his spiritual experiences an extremely concrete character, and it also indicates his evaluation of contemporary generation theories, for example, Swammerdam's preformation theory.

Swedenborg's contributions in the area of brain research are among the best known of his scientific works, and it has been claimed that his theories on the location of the mental processes in the cerebral cortex were far ahead of his time and, in part, deserve high praise.[37] However, what little he himself published of his brain studies drowned in the mass of more sensational material, and it was not until Rudolph L. Tafel published a kind of anthology of manuscripts in the 1880s that it became clear how far-sighted Swedenborg had been.[38] It should nevertheless be stressed that, in this field also, he adhered to the method mentioned earlier: he applied his brain to what others had worked on with their hands; thus, he collected an enormous material from the leading anatomists and drew his own conclusions from it.

Swedenborg himself did not publish any brain studies other than the introductory chapters to Part II of *Oeconomia* (1740–1741). The motif for the entire work, which was based on investigations of the composition of the blood, is presented in the first sentence of the first volume:

> The animal kingdom, the economy of which I am about to consider anatomically, physically, and philosophically, regards the blood as its common fountain and general principle. In undertaking, therefore, to treat of this economy, the doctrine of the blood must be the first propounded, although it is the last that is capable of being brought to completion.[39]

This is quite logical in view of the emphasis on the elements' movements in the body, which appeared in the treatise on the mechanism of soul and body in 1734. Swedenborg now believed he could distinguish three different degrees of blood, from ordinary red blood via a purer form to the highest, which is identical with the spirit of life, and he based his views mainly on Leeuwenhoek's microscopic discoveries of the blood corpuscles.

This classification created entirely new possibilities for weaving the physiological and mental processes into an all-inclusive circulation, which appears not least clearly from a glance at the description of the cerebral cortex, the point of departure for the study of the brain in *Oeconomia.* In the very first manuscripts on the physiology of the brain, roughly dated by Alfred Acton at 1738, Swedenborg concluded from his sources that the cortex is the noblest part of the brain and that it consists of innumerable small parts held together by the finest fibers emanating from the outermost arterial vessels in the pia meninx, the vascular membrane that lies closest to the surface of the brain.[40] Through the arteries in this membrane flows the purest blood, the spirit of life, into the cortical substance, and from there it moves on to the fibers that form the medullar parts of the brain, the spiral marrow, and the nervous system.

The cortical substance thus came to occupy an intermediary position, and this caused Swedenborg to revive the same mythical symbol which he used in a similar context in the *Principia*, the two-faced Janus:

> Therefore the cortical substance is placed in the middle, or in the last term of the arteries of the brain, and in the first term of the fibres of the brain, so that this substance, like the two-headed Janus, looks backwards and forwards; backwards on the side of the arteries to the crasser blood, but forwards on the side of the fibres to the spirituous fluid, both of which unite in a manner in the cortex as their common and principal substance; and this, in order that effects may return to their causes in a wonderful circle every time that causes tend to their effects; and *vice versa*.[41]

As mentioned earlier, the Janus symbol was used in *Principia* to indicate the boundary line between infinite and finite, the mathematical point. Here it occupies the same place and marks the border between physical and psychical in the body of man. But in contrast to the mathematical point, the cortex of the brain is exceedingly complex. Its smallest parts, the brain cells—*cerebellula* or *sphaerulae*, to borrow Swedenborg's own terms—are admittedly products of branches of blood vessels, although they cannot be regarded as vessels themselves but rather as a kind of membrane enclosed in spheres. Once more we find ourselves very close to the mechanistic model of 1734.

Armed with microscopes, Leeuwenhoek and Malpighi had already reached this point, but Swedenborg's imagination would not let him rest there. He conceived of cells of even more advanced and more perfect form, comprising a kind of *cortex cerebelluli*, a cerebral cortex of the brain cell. Each one of these brain cells he regarded as a miniature brain, which gives rise to an indefinite number of "brain particles," in the same way as the physical particles are indefinitely divisible in *Principia*. To make these subtle materials more concrete, let us look ahead to the anatomical demonstration which Swedenborg let an angel make, when he gave his imagination free rein and poetic shape in *De Cultu et Amore Dei* in 1745. The angel begins with a nerve that it exposes to show how it consists of an endless mass of fibers; one of them, magnified by the angel's gift which exceeds all mundane lenses in acuity, is studied closely. Eve is then allowed to follow the course of this fiber to its origin in the cortex of the brain, whereupon begins the demonstration of the structure of the brain:

> Having withdrawn therefore the softest membranes, the first object presented to view was a kind of new brain, but in a diminutive form, again with infinite spherules, or little spheres, arranged into the infra-celestial form, all of which had a fixed relation and view to greater and lesser circles, and to their poles altogether as in the great sphere of the world. It was also rendered visible, in what manner this form taken from little spheres, by the variations of itself and changes of state, produced ideas called material; and in what manner each little sphere sent forth a diminutive fibre with its little duct and covered it with a small coat; and how natural life was infused into it from the lowest spiritual fountain which inhabits that sphere with its genii, and excites its organic principles. . . .
>
> Having examined these things, she next unclosed and opened one of these little spheres, and inwardly in it she again brought forth to view innumerable new little vortices, the highly adorned dwellings, as it were, of so many intelligences and wisdoms winding into a celestial form . . . from stamina as numerous as were the little vortices or the small habitations of the intelligences, /they/ contrived the superficies of the above-mentioned fibril, which is permeable to natural life.
>
> Again, one of these little vortices or little stars being laid open, there appeared the supreme of all forms, called the super-celestial, from which darted those rays, or fibres by supereminence, which being permeable to the life of the Supreme penetrated into Olympus.[42]

It is in these terms, which in unique fashion combine a passion for detail and precision with profound reverence and tenderness, that Swedenborg tried to capture his fleeting vision of the sources of psychical life in the human brain. But here he went further than in his scientific works, primarily by introducing divine love as an active force in the cerebral passages. A consistent characteristic of his writings after the crisis of his conversion is that he describes the world of spirits in human biological, not least cerebro-physiological terms; as early as in *Arcana Coelestia,* heaven is seen as a great human being in whom all organs are represented through innumerable spiritual societies. A general inspiration for this vision may have been Plato's *The Republic*; of greater importance, obviously, are the decades of study and speculation centered on the most subtle structures of the organism. By combining his experiences of these years with the most audacious analogical reasoning, Swedenborg's imagination finally succeeded in bridging over the Cartesian abyss between matter and spirit.

Nevertheless, when the time came to edit *Oeconomia* at the end of the 1730s, Swedenborg still had a long road to travel. This extensive work has in no measure received the attention it deserves from the scholarly community, but it has much to offer, not least on the problem of the spirits of life. In the mechanistic essay of 1734, we saw that Swedenborg was striving to get away from a concept which he found inexact and "occult," and to replace it by a realistic, empirically founded understanding of the movements of the elements in the body. It might seem that he himself was now guilty of the same occultism. After all, he presents here a rather definite hypothesis to the effect that the spiritual fluid, which appears to be identical with the spirit of life, is produced by the innermost degree of brain cells and flows out into the organism through two channels, partly via the nerves and partly in combination with the blood. The theory is new to the extent that it presupposes a more universal circulation than that of the blood, and it is thus unquestionably more sophisticated than Descartes' concept of the epiphysis as an exchange center for information between soul and body. Above all, it provides some sort of an explanation of the vital function of the cerebral cortex. But it does not give an answer to the question of how spirituality can be combined with an organic fluid, and this, of course, is the crucial problem.

Nor did Swedenborg believe that this theory provided the answer. He still did not master the language that would make possible the all-inclusive formulations required to solve the problem. Meanwhile, he tried to associate various bodily fluids with nature's elements in the same way in principle as in the sketch of the soul of 1734. The cells of the brain undergo changes according to environmental variations, that is, primarily according to the condition of the blood, and this means that they adapt themselves to the auras or elements of nature. The ear is influenced by modulations in the air, and the eye adapts itself to changes in the ether. Hence, the same must be true of the inner sensory organs in relation to the higher auras in nature, which have a much greater ability to penetrate tissues than the physical media of sound and light.

Characteristically, Swedenborg looked for support to two great scholars, Swammerdam, who described the air passages in certain insects in a way that Swedenborg could associate with blood vessels in higher beings, and Leeuwenhoek, who showed that the movements of the blood corpuscles can be influenced by air currents. It was on this foundation that he constructed his analogical conclusion: red blood represents air in the organism, the purer blood the ether. The spiritual fluid must then be affected by the highest auras of nature, in animals by the next highest, that is, the magnetic, which, as already said, would explain their sense of direction, and in man by the highest.

It seems clear that we must not take this to mean that Swedenborg believed in an automatic influence by the natural elements on man. This is inconceivable by the very fact that the brain itself can change the trends of its reactions, which is a physiological way of establishing freedom of will, something that Swedenborg never wished or dared to question. But, in addition, the influence of the auras is modified to suit the specific qualities of the organism, the most important of which, of course, is life. These modifications include heat in nature equivalent to anger in man, cold to fear, expansion to pride, motion to action.[43]

Obviously it became more difficult for Swedenborg to establish these connections the further he reached in psychical complexities, but here he repeatedly refers to a sort of mathematical universal language that could formulate psychical and physical phenomena with the help of the same system of symbols. Until someone succeeds in creating such a language, we are compelled to resort to inexact comparisons, such as that between the spirit of life and nature's most exalted aura. This is the background against which we must judge Swedenborg's use of the notion of the spirit of life. He hesitated to abandon his very definite formulations of 1734, but he was compelled to undergo the same experiences as all other scholars when faced with the transition from the planning of a great research project to its execution: problems prove to be much more difficult than one could ever anticipate, and our knowledge is and remains fragmentary. In the famous remarks of the old Sir Isaac Newton, the great idol of the Enlightenment: "I do not know what I may appear to the world, but to myself I seem to have been only like a boy playing on the seashore, and diverting myself in now and then finding a smoother pebble or a prettier shell than ordinary, whilst the great ocean of truth lay all undiscovered before me."[44]

The Concept of the Formative Force

We shall return to the problem of a universal language later on, but before we do so other biological phenomena must be discussed. As mentioned in the introductory survey, Cartesian mechanistic biology was especially vulnerable when it came to explaining the origin of new living organisms. It is thus quite natural that Swedenborg exceeded his Cartesian limitations in the section of *Oeconomia* concerning the genesis of an organism. This section treats one of the favorite topics of contemporary biology and proceeds from the work of the great Italian anatomist Malpighi issued in 1672 under the same title as Swedenborg's chapter, *De formatione pulli in ovo* (On the Formation of the Chick in the Egg).

The real purpose was to prove that it is the brain that is first formed in the egg, while the heart and the blood vessels come next. The consequences are important enough in themselves, but even more significant is another principle advanced quite unexpectedly, namely, that a formative force, *vis formatrix*, is present in the organism. Since Descartes's great contribution was to eliminate all scholastic "forces," all those occult *facultates*, from the realm of geometry he so scrupulously sterilized, Swedenborg's thesis may appear as a break with his entire Cartesian past:

> There is a certain formative substance or force, that draws the thread from the first living point, and afterwards continues it to the last point of life. This is called by some the plastic force, and the Archaeus; by others, simply nature in action; but I think it will be more intelligible if in reference to the work of formation we term it the formative force and substance.[45]

Swedenborg did not closely define this force, however; and when he refers to it once more in the drama of creation *De Cultu et Amore Dei* in 1745 in connection with the genesis of plants, he again evaded his obligation to give a definition:

> The first generating or plastic force, innate in the very seeds of vegetable foetuses, may be likened to a soul. . . . But what the quality of this seminal force is, cannot be known but by unfolding the forms of prior nature, both those which have reference to active and those which have reference to passive powers, also in what manner the solar rays operate to join them together, to evolve which, from lasts to firsts, would be too vast an undertaking.[46]

There is, however, a highly respectable reason for this unwillingness to spell out a definition, which is also significant in evaluating the influence of Descartes. A strict definition simply cannot be achieved with the language at our disposal, since the phenomenon is beyond the sphere of words; once more what is needed is the artificial philosophical-mathematical language for which Swedenborg so frequently expressed a need during his last decade as a natural scientist. It is important to bear in mind that this desire for a mathematical language, which he shared with so many of his predecessors, had one of its roots in Descartes. For this reason alone, Swedenborg need not have regarded his theory of the formative substance as a radical break with the past, but rather as a new attempt to arrive at a more subtle, mechanistic system than his teacher's. His choice of terminology also points in that direction. He speaks of a formative force, *vis formatrix*, and by "formative" he means something highly concrete, *gestalt* and figuration in the organs he was studying. It is another matter that, ever since Aristotle, the concept of form has been so fraught with metaphysical import that it led his thoughts to other domains than the geometrical.

The other terms Swedenborg mentioned, *vis plastica, Archaeus, natura agens*, caused certain scholars to draw conclusions concerning his affiliation with a "mystical" philosophy of nature of ancient vintage. *Archaeus* belongs to Paracelsus (1493–1541) and John Baptiste van Helmont (1577–1644) and thus is associated with Renaissance medicine and the iatrochemical principles of the seventeenth century. *Vis plastica* is a term belonging to the Cambridge Platonists, and recalls Cudworth's and Henry More's aspirations in the England of the end of the seventeenth century to combine Christianity and philosophy into a kind of modern Platonism. Listing the terms, however, tells nothing of the more profound influences. It is probable that Swedenborg acquired his knowledge in some handbook or another, e.g., Johann Christopher Sturm's *Physica electiva sive hypothetica* (1697), a long and eclectic textbook in physics, where he could acquire broad knowledge of the older as well as the more recent philosophy of nature; all the terms and sources to which he refers are thoroughly reviewed and analyzed in it.[47]

It is less rewarding to attempt to put Swedenborg in a framework of "mystical" philosophy of nature with direct connections to the Renaissance than to examine his relationship with the intellectual dialogue of the times. *Oeconomia Regni Animalis* was prepared during his years of foreign travel and begun in Paris. Swedenborg found the intellectual environment of France familiar and attractive in many respects. It is true that Descartes had been severely criticized on important points, though not more severely than by himself, but the Cartesian rationalistic methodology was nevertheless still the foundation of French natural science. The Anglophile Voltaire complained at this very time about what he considered the backwardness of French science, both in his banned and burned *Lettres philosophiques* and in the tolerated but controversial *Eléments de la Philosophie de Newton.* At least partly as a physicist and definitely as a philosopher Swedenborg corresponded well to Voltaire's description in the work on Newton, an immensely influential treatise, one of the fundamental works of the Enlightenment:

> No good physicist exists today who does not acknowledge both Kepler's rule and the necessity to acknowledge a gravitation as demonstrated by Newton; but there are still philosophers who cling to their vortices of subtle matter, who desire to combine these imaginary vortices with these proven truths.[48]

The Cartesian vortices, as we know, still had a central function in Swedenborg's philosophy of nature, and they recur in his doctrine of form and psychophysical speculations, as we shall see later on.

Voltaire contrasted such representatives of seventeenth-century rationalism as Malebranche and Leibniz with his English ideals, Newton and Locke. (I shall return later to Locke's significance to Swedenborg.) From this distance, both Malebranche and Leibniz appear to have been more important than most as sources of inspiration for Swedenborg. True, he had known of them earlier via different handbooks, but he appeared not to have begun a systematic study of their works until after his stay in France. Considerable space is devoted to both of them in the philosophical and psychological excerpts, which were collected in the manuscript 36-110 beginning about 1740.[49]

It cannot be proved that it was Voltaire's books written in England that stimulated these studies—scarcely anything can be proved about Swedenborg's activities in Paris, except that earlier allegations about his studies at l'Ecole de Chirurgie are completely unfounded.[50] On the other hand, the intellectual milieu unquestionably provided numerous examples showing that the rationalism of Malebranche and Leibniz was still alive, not least when applied to biology. In his great study on the influence of the biological sciences on French thought in the eighteenth century, Jacques Roger repeatedly refers to these two great men and their relations with the natural scientists:

> For Malebranche as for Leibniz, God subjected Himself to order and did not act without an adequate foundation, and this foundation,

> this order, though it infinitely surpassed human comprehension, was nevertheless not entirely inaccessible to mankind. Great scientists like Malpighi, Tournefort, Vallisneri, and Réaumur believed in a Godgiven order which science can come to understand.[51]

This is the tradition to which Swedenborg belongs. It might appear antiquated in a situation where the deistic attacks from England were becoming increasingly violent, with their demands that natural science should confine itself to describing and formulating laws as accurately as possible and forget its ambition to explain the most profound causes, Newton's statement hypotheses *non fingo* being the leading model. At the same time, it is important to recall that Leibniz experienced a sort of renaissance in France during the years immediately following 1735, due in part to the spread of Wolffianism in Germany; and even those who rejected Leibniz's metaphysics retained his monad concept in both organic and inorganic contexts as a basis for an atomistic approach to nature. Roger even speaks of a "néo-Leibnizianisme" during this period, that is, when Swedenborg was living in Paris.[52] Thus we have every reason to study Leibniz's influence in *Oeconomia*, which means that we can pick up the threads from the beginning of this chapter, beginning with the genesis of the organisms.

Swedenborg quoted Leibniz's theodicy in the Latin edition of 1739. In the preface, Leibniz refers to Cudworth's *vis plastica* in a controversial sense, asserting that he could not find it necessary to resort to special formative powers to explain the origin of organisms. It can be explained to complete satisfaction within the framework of a mechanistic system, assuming that one accepts pre-existence in seed and eggs. But the embryological theory is placed in a greater context, which is the essential one:

> To explain this miracle, the formation of organisms, I employ a predetermined harmony, that is to say, the same means I used to explain another miracle, the correspondence between soul and body, where I proved the uniformity and fertility of the principles I myself have used.[53]

As is well known, the question of the relationship between soul and body became a problem of crucial importance after Descartes and was of constant concern to Swedenborg. In Leibniz, Swedenborg discovered an attempt to solve it together with the problem of the origin of organisms within the framework of a single intellectual model, namely, the concept of a *harmonia praestabilita*, a predetermined harmony.

At first glance, it might seem unjustified to draw a parallel with Leibniz, since Swedenborg flatly rejected the idea of preexistence, the idea that an infinitesimal model of the complete organism is present in the egg, which simply increases in size. Despite the fact that he proceeded from Malpighi, he seems to have adopted the opposite opinion, according to which the organism develops from a completely chaotic mass in the egg, a notion that is often attributed to Aristotle and that was shared by both Harvey and Descartes.[54] However, the essential point is not this technical difference, which is of subordinate importance, but rather the idea of the concept of force, the force that Swedenborg identified with the soul in organic contexts. Here the correspondence with Leibniz is striking.

In *Système nouveau de la Nature et de la Communication des Substances* (1695), Leibniz described the development of his thinking. He tells how, after he had discarded Aristotle, he became engrossed in ideas of empty space and atoms. But these ideas did not suffice to explain the existence of unity; the principles of unity could not be traced exclusively to matter, to whatever is passive. Nothing then remained but to resuscitate the substantial forms so maligned at the time, which led to a pronounced Aristotelian renaissance in the Leibniz tradition. Leibniz found the true essence of these forms in the force concept, which was regarded as related to human feelings and passions, hence analogous with the soul concept. But the soul cannot be used to explain every detail of the utilitarian construction of the body, nor can the substantial forms be used in analyzing each individual problem in nature. Instead they simply comprise the essential prerequisite for the creation of true general principles, while the rest of nature must be studied from mechanistic points of departure.[55] So far the parallel with Swedenborg could scarcely be more perfect.

These forms and souls would have to be indivisible, like the human soul, but this posed new and difficult problems. Every substance that comprises a true unity can neither come into existence nor cease to exist except by a miracle, and the forms must therefore be the result of an act of creation, in the same way as Genesis taught us about the human soul. It was precisely in this connection that Swammerdam's, Malpighi's, and Leeuwenhoek's observations were so useful to Leibniz, since he saw in them the opportunity to interpret that which appeared to be the birth of an organism not as a truly new creation but as a transformation of an organism created in the beginning. As has been pointed out, Leibniz, by regarding seed as possessing a kind of soul from the beginning, constructed a biological entity equipped with its specific energy. Thereby he achieved an outline of a dynamic theory, which, in the long run, defeated the mechanistically rigid system he was trying to save.

This is especially noticeable in his treatment of Leeuwenhoek's animalculist theory. (Swedenborg's knowledge of this and other contemporary theories of procreation is clearly revealed in many of his works.) The discoverer himself had assumed that the spermatozoa contained a motor principle, or a kind of soul, from the creation, and they therefore represented in Leibniz's eyes the perfect example of those bodies with which every soul—or every monad, to use a more specifically Leibnizian term—is constantly united and which themselves are subject to the conditions of constant change. At the moment of conception, each animalculum takes on a new shape, which allows it to absorb nourishment and to increase in size and, after the fetal stage, to emerge into a new and greater arena.

To sidestep Leeuwenhoek's difficulty in explaining how humans can develop from diminutive animalcula, Leibniz tried one or two lines of reasoning. According to one, the souls possessed by human spermatozoa have no intelligence until after fertilization, but this is a transformation that requires divine intervention. He also entertained another notion, namely, that only the spermatozoa intended to form human beings possess human souls, which would explain why all existing souls are involved with original sin, at the same time as it eliminates the need for constant divine intervention at the stage of conception. Both solutions, nevertheless, have the same favorable result in one important respect: they dispense with the troublesome idea of a mass death of human *Anlage,* since those "worms" that do not develop into beings remain worms and nothing but worms.

This result is in line with a basic tendency in Leibniz's entire idealism: there is actually neither birth nor death in nature but only transformations and changes in forms of manifestation.[56] The parallel with Swedenborg is remarkable in this respect also, in fact so close that we have reason to see his final vision of spiritual societies concentrated in a universal human figure as a concretion of Leibniz's monad theory, as a realization of the uttermost consequences of the system of pre-established harmony in the imagination of a poet and prophet.

Leibniz could thus make excellent use of Leeuwenhoek's microscopic discoveries in developing his thinking, but he was not inclined to harness his philosophy to a scientific theory that could be disproved by later research. In the long run, it turned out that Leibniz's philosophy led away from the pre-existence theories, both the ovist and the animalculist version, and toward a dynamic point of view. It has been questioned whether Leibniz was on the verge of such a radically new concept as evolution.[57] However, the texts that would justify this speculation were not found until 1765, almost fifty years after Leibniz's death, when *Nouveaux Essais sur l'Entendement humain* (New Essays concerning Human Understanding) were published. The result was a renaissance for Leibniz, this, again, being an important factor in the genesis of the philosophy of German Romanticism. From our point of view, it is essential to note that the innate dynamism of Leibniz's system had long before that time belonged to the essence of Swedenborg's biological thinking, and it occasionally seems as if too little attention has been paid to Swedenborg as a link between seventeenth-century rationalism and the natural philosophy of Romanticism. In any case, it is unquestionable that he further advanced Leibniz's dynamistic ideas.

The great vision Leibniz tried to capture in his monad model first emerged in *Nouveaux Essais.* Not until then could one visualize the sequence of autonomous psychic entities, each one, according to its ability, a reflection of the entire universe—a concept that allowed Leibniz to transform the Cartesian machine into a spiritual organism. In the period between Leibniz's death in 1716 and the publication of *Nouveaux Essais,* the principal advocate of Leibnizianism was Christian Wolff (1679–1754), the prototype of the solid German scholar, who, in a long series of handbooks, reduced Leibniz's constantly evolving philosophy to a rigid system that could be taught to large numbers of students in the classroom. At least this is the usual view of Wolff, spiced with the absurd optimism of Voltaire's caricature, Doctor Pangloss in *Candide.*

But like all popular evaluations, this too is one-sided and misleading, and even Voltaire had expressed his appreciation of Wolff in earlier works: in the famous biography of Charles XII, for example, he called him "that renowned and eminent connoisseur of all branches of philosophy."[58] In Swedenborg's time, Wolff was regarded both as a preacher of dangerous heresies—in 1723 he was expelled from the university in Halle by the pietistic faculty of theology, which was as adamantly opposed to freedom of thought as its orthodox foes—and as a bold pioneer, the formulator of a highly attractive modern rationalism. As has been shown by Tore Frängsmyr, Wolff's philosophy was regarded with great suspicion by the authorities in Sweden in the beginning of the 1730s, but at the end of the decade it came into favor instead.[59]

Even if he thus had to be rather cautious in public, both in *Principia* and *De Infinito,* Swedenborg became fascinated by Wolff and his psychological, ontological, and cosmological series of handbooks in the early 1730s. It is true that he later changed his opinion—in certain *memorabilia* from the spiritual world, Wolff was pictured in humiliating situations and always subordinate to Leibniz; still, there is no doubt that Wolff was of great importance to Swedenborg, particularly in his psychophysical thinking during his last decade as a scientist. According to his Latin travel diary, he had begun to read

Wolff in Dresden in July 1733, and the experience was so overwhelming that he wrote an appendix to *Principia*, in which he took a stand on Wolff's principles and was eager to emphasize both the almost complete correspondences and his own independence. He explains that he had formulated his own theses two years before he began to study Wolff.[60] Quotations from Wolff occupy a great deal of space in the manuscripts, a matter to which we shall return later in another context.

Haec vera sunt quia signum habeo

No matter how much remains to be discussed in Swedenborg's biological researches, we must turn to an examination of their psychological consequences, first, though, a quick insight into Swedenborg's working methods, which a volume of manuscripts can give us. He finished the first part of *Oeconomia* on December 27, 1739, in Amsterdam. During the first months of 1740, he was busy with the preparations for the second part, beginning with the dura mater, the hard membrane of the brain. Since he noted dates, we can observe his diligence step by step and find it extremely impressive. At the beginning of February, however, he stopped his own writing to quote a number of anatomists: Morgagni, Boerhaave, Ridley, Winslow, and Leeuwenhoek, among others. Scarcely had he begun his notes on Leeuwenhoek's *Arcana naturae detecta* (Nature's Secrets Revealed) than he suddenly broke off and wrote a two-page summary of his own corpuscular philosophy; after that the quotations begin anew.[61]

This summary is entitled *Philosophia corpuscularis in compendio* (A Compendium of Corpuscular Philosophy) and has often been cited because it ends with a strange and prophetic sentence. However, due to incomplete publication of the manuscript, scarcely anyone has observed its framework, which is essential for a correct interpretation. The hasty notes briefly outline the series of particles in nature. They begin with the world's first substance, mention the element of fire and then the four auras of nature from which the spiritual fluids in the organisms are determined. Then come the terrestrial particles, first the round ones that form metals and water, and then the angular *corpuscula* that are found in salts, acids, and alkalies. Oils and various kinds of volatile essences constitute the subdeterminant class, which produces a number of different salts.

The whole is a general chart of the elementary chemical structures of nature, which is in line with Swedenborg's views expressed in his draft papers on the philosophy of nature twenty years previously; furthermore, he relates spiritual fluids to the highest elements. However, he was disinclined to go into detail at this point:

> Nevertheless, by these determinants, by means of subdeterminants, compositions of infinite number can be formed, especially in the vegetable kingdom; thus bullulae or vesicles, having at length become fixed, broken up, the aura flying away, or otherwise compressed, produce essential juices, and all the flavor in juices, etc. These things are true because I have the sign.[62]

It may seem strange to encounter this general declaration in the midst of a series of anatomical excerpts, but the explanation is to be found in Leeuwenhoek's text. Swedenborg first jotted down a few passages, in which the microscopist described how infinitesimal salt particles retain the same square shape no matter how finely they are divided. Leeuwenhoek continued with a rapturous description of the *animalcula* that are to be found in ordinary water and that, even though they are no larger than grains of salt, have feet and legs and entrails and blood vessels. If he could dissect these bodies, Leeuwenhoek was convinced that he would find the same salt particles in their blood as in man. Swedenborg copied the last sentence in the paragraph, correctly for once—normally he wrote too quickly to bother about the accuracy of quotations:

> I also do not doubt that such a large amount of salt particles is spread over our whole body that each and every blood corpuscle consists of many salt particles.[63]

Leeuwenhoek devotes two or three more pages to a discussion of the occurrence and forms of these particles and finally concludes that his views conflict with those of "the brilliant René Descartes" in certain respects.

In all probability this is the point at which Swedenborg broke off his reading in order to summarize his own particle doctrine, which in principle retained the stamp of Cartesianism. He felt himself called upon to clarify his own stand by reading Leeuwenhoek's self-asssured and somewhat aggressive presentation, and this makes the oft-quoted final lines more understandable, at least on one point. Leeuwenhoek frequently expresses himself hypothetically and uses the formula *si verum est* (if this is true), and one has the feeling that Swedenborg replied to the dead microscopist: *haec vera sunt quia signum habeo* (these things are true because I have the sign).

The meaning of this reference to "signs" will be discussed later, but the episode itself is a thrilling presage of what will be so common in the descriptions of the spiritual world contained in the theosophical writings. We find Swedenborg involved in a posthumous debate with one of the greatest natural scientists of the epoch, incidentally in the very city of the interlocutor's youth, and he finds confirmation of his own views through a sign, just as his arguments with various philosophers in the spiritual world will be decided in his favor through various signs from the heavens. To the solitary, intensely industrious natural scientist then residing in Amsterdam, his scientific conviction seems to be taking on the form of a revelation. But the connections with the mechanistic thinking of his youth still proved to be utterly valid, and tremendous efforts still had to be made before nature's secrets were rejected in favor of heaven's, Leeuwenhoek's *arcana naturae* for his own *arcana coelestia.*

4

Ontology and Psychology

Among the stumbling blocks for the reader of Swedenborg is the man's synthetical aspirations, which find universal contexts in each small problem, and his constantly shifting aspects. Both factors make it difficult to discern clear lines of development and to establish sources, but at the same time make reading his works more exciting. During the decade before 1745, Swedenborg's thinking was crammed with vital dynamism and the joy of learning, combined with a search for comprehensive visions to an extent that is rare in European intellectual history.

The Doctrine of Series and Degrees

Despite his eclecticism, Swedenborg on several occasions followed closely in the wake of contemporary "synthesis makers," although never losing his critical sense. This tendency becomes obvious if we pause to consider the central problem in the Cartesian tradition, the relations between soul and body. As late as 1769, Swedenborg discussed the various theories for the solution of the problem in a small treatise called *De Commercio Animae et Corporis* (The Intercourse between the Soul and the Body). The first time he took up the question was in the final chapter of *Oeconomia* I, entitled "Introduction to Rational Psychology," and it is by no means only the title that recalls Wolff's writings.

Swedenborg presents three hypotheses to explain the relationship between soul and body:

(1) the traditional view as it was developed by Aristotle and Scholasticism, which asserts that there is a direct exchange of bodily reports and spiritual orders; Swedenborg employs the usual term, *influxus physicus* (bodily inflow).

(2) Occasionalism, the notion that each transfer from spiritual to physical and vice versa can only be explained by divine intervention; Swedenborg ascribed this view to Descartes, as was usual at the time, although it was actually elaborated by his disciples Guelincx and Malebranche and others.

(3) The *harmonia praestabilita* theory, Leibniz–Wolff's idea to the effect that there is actually no relationship between soul and body, but instead that predetermined parallel events occur in such a way as to give the impression of relationships. The psychophysical parallel theory developed by Spinoza is never mentioned by Swedenborg, and its pantheistic motivation was naturally inconceivable to him, as it was to his contemporaries. His own final solution, however, closely resembled a parallel theory, as we shall see later.

In *Oeconomia*, however, Swedenborg was dissatisfied with all three theories, considering that they assumed that the soul immediately flows into the sphere of bodily effects, whether this originates in its own inherent principles or in some superior ones, and he found that this would be to draw premature conclusions.[64] What was needed was a more profound understanding of the order of life. and before he provided his own solution he tried to formulate a doctrine of the coherence of things. He calls this the theory of series and degrees: later he defines it with the doctrine for which his name became best known, namely, the theory of correspondence.

According to Swedenborg, there is nothing in the entire universe that does not belong to a series or represent a degree in a series, with the exception of the first substance of nature. In direct connection with *clarissimus Wolffius*, the renowned Wolff, Swedenborg explained that the most extensive of these series naturally consists of the natural universe itself. Within this universe there are, in turn, six great series, three higher and three lower. The first of the high series originates in the first substance of nature. The second consists of the substances of the first—Swedenborg uses the term "substance" at several levels, which scarcely contributes to clarity—when they have been released to gyrate in Cartesian movements, which gives rise to the element fire. And the third comprises the four auras of nature. The three lower series consist of the three kingdoms of nature, mineral, vegetable, and animal.[65] Each one of these great series consists, in turn, of a multitude of subordinate series, just as the cerebral cortex constitutes a series made up of an indefinite number of brain cell series. It is not difficult to recognize Swedenborg's pattern from Leeuwenhoek's attempts to explain what he saw in the microscope.

Thus, the only exception to this universal affiliation with series is the first substance of nature. However, as was first asserted by Aristotle, we can have no certain understanding of this but can draw only some analogical conclusions. Swedenborg also quickly dropped this problem and instead concentrated on the first substance in each series, that is, "substance" in a transferred sense. All series have a first substance of this kind, as many scholars have established, and terms like *monad* and *simple substance* have been used to define it. But Swedenborg disapproved of all earlier terminology, about which he had learned partly from Wolff's textbooks in ontology and cosmology and partly from a pre-Cartesian handbook of 1636, for the reason that all terms that try to suggest simplicity for these first substances give the inaccurate and dangerous impression that they originate directly in the first substance of nature.[66] In this way, one could be misled to believe that all the imperfections and flaws in nature are directly caused by nature's first substance or were contained in it at creation.

If we recall how Leibniz imagined that the souls or monads had been in existence ever since creation and combine this with his concept of the world as the best of all conceivable worlds, it is difficult not to acknowledge that Swedenborg's arguments had some validity. Obviously, he himself experienced one of the great problems inherent in the idea of pre-established harmony at this very point, and he tried in various ways to evade the static, the pre-determined in the Leibniz–Wolff model. In *Oeconomia*, Swedenborg suggests that *harmonia praestabilita* be replaced by what he calls *harmonia constabilita*, a successively developing harmony, in which the substance of each series, according to decreasing capacity, tries to carry out the intentions of the first substance.[67] Here we see most distinctly Swedenborg's strong dynamism, at the same time as his close kinship with Leibniz is very apparent.

To some extent, naturally, the contrast with Leibniz is illusory, and it can probably be explained largely by the fact that Swedenborg proceeded from Leibnizianism in the rather scholastic form Wolff gave it. Swedenborg also must have believed that the highest form of harmony was predestined in a specific phenomenon, namely, in the first substance of nature, which is beyond and above the entire system of series and degrees in the Creator's first realization of himself in his nature. In biological contexts, Leibniz, in turn, had the same idea as Swammerdam—that original sin could be interpreted as a concrete fact within organic nature. It is indisputable that Swedenborg tried hard to learn Wolff's system and terminology, but it is equally clear that he did not want to do so slavishly; his irritation is unmistakable on many occasions.

The theory of series and degrees was developed in close relationship with mathematical terminology, and Swedenborg clearly proceeded from Wolff's definition of the concept of series and degrees, although at times he went further in mathematical concretion than the German philosopher. For example, Wolff refrained from using Leibniz's term *monad* when referring to the simple elements of the body, probably because he wanted to avoid the mathematical associations of the word and of its Latin translation, *unitas*, unit. But Swedenborg often used the word *unitas* for a degree in a series, because he could then compare things comprised of such degrees with mathematical figures, homogeneous and comprised of units.[68] This train of thought was admittedly never carried beyond a purely theoretical plane, not for any lack of trying, but because Swedenborg never managed to complete the project to which this terminology was a preliminary, that is, an all-inclusive, universal system of mathematics. Even at its preparatory stage, his struggle with a mathematical language may have helped to lay the ground for the symbolization of reality, which slowly took shape during this period. It is noteworthy that Swedenborg's terminological observations end in a reminder of the ancient scholar for whom the world of figures became the world of mysticism: "The Pythagorean philosophy seems to have acknowledged similar units, having their harmonies and concords, which it compares with the units of numbers."[69]

As he moved toward ever higher levels of complexity, Swedenborg constantly sought support both in the wisdom of the ancients and in the sciences based on modern experience. In the struggle between *les anciens* and *les modernes*—the debate over many decades that in French intellectual history is usually called *la querelle*, "the quarrel"—Swedenborg thus took an intermediary stand, which is also apparent in his treatment of the soul concept. As we have observed his way of handling it so far, it is a relatively simple and uniform concept: the soul in contrast to the body, the soul in coordination with the body, the soul as a spiritual being—all this sounds fairly comprehensible and intelligible. Contemporary authorities, who were still his mentors, were also content with relatively simple concepts, although there were of course exceptions.

The situation changed when Swedenborg began to penetrate the problems of the soul in greater depth. Now he needed a much more sensitive and precise conceptual battery. Swedenborg himself indicated that the final chapter of *Oeconomia Regni Animalis* II, entitled "On the Human Soul," represented the first fruits of his psychological endeavors; but even in the first part of *Oeconomia*, he had reached the point of identifying the soul fluid with the formative substance of the organism. In contrast with other parts of his writings, he is generous with proofs that make it possible to establish his sources. Nevertheless, we may have the erroneous impression of tremendous erudition, unless we remember that he usually knew his sources only indirectly, through handbooks and works by contemporary scientists.[70]

In *Oeconomia* II, the point of departure also was the spiritual fluid in the organism, the receptive organ for inflowing life which, Swedenborg asserted in the embryological chapters of the first volume, comprises the formative substance of the body. By introducing the concept of form in his system, Swedenborg not only affiliated himself with Leibniz and Wolff but also came into intimate contact with the older scholastic philosophy and its subtle speculations about this concept. It is also indisputable that he made use of the form concept in a manner that recalls Scholasticism. He let it glide from the forms of concrete objects over such abstractions as social form and thought form to ideas and representations. In a posthumously published essay on the fibers of the body, which was intended to constitute part III of the *Oeconomia* series, he expounded a doctrine of form that clearly reveals his tendency for hypostatic thinking; we shall return to this doctrine at the end of the chapter.

When in *Oeconomia* II he calls the spiritual fluid the form of forms, Swedenborg is nevertheless careful not to confuse it with what Aristotle called pure form—a kind of god concept—but stresses that the term should be regarded as a parallel to what Wolff called a representation of the universe. This obviously means that the *fluidum spirituosum* includes things that we cannot comprehend rationally. The spiritual fluid seems further to be identical with what Cartesian psychology called *spiritus animalis*, the spirit of life, but Swedenborg now tried a more daring approach: he identified *fluidum spirituosum* with the soul itself, with *anima*, to use his own Latin term. This is the most extreme expression of his original aspiration to regard the soul as a finite being, capable of being penetrated by mechanistic methods. In its physiological manifestation, the spiritual fluid could thus be captured for analysis in a future which would have microscopes more powerful than Leeuwenhoek's.

To understand Swedenborg's later development, we must obviously retain this image of the soul as a constantly pulsating fluid in the organism. At the same time, it is clear that the soul's physiological function is only one side of this Janus visage. In its corresponding psychological gestalt, the spiritual fluid represents the highest form of thought and emotion, a medium for life and wisdom as they flow from the divine source. Swedenborg makes a few pious observations on this in the last chapter of *Oeconomia*, although still without crossing the line into theological domains. Of greatest importance, however, is the fact that during his work on ontological problems after 1736—a study conducted primarily with the help of Wolff's *Ontologia* and two older handbooks, Scipion Dupleix's *Corps de Philosophie* (1636) and Robert Baronius's *Metaphysica generalis*, which provided Swedenborg with practically all his references to classical antiquity—he came increasingly to dissolve the boundaries between matter and spirit.[71] Identifying the soul with the organic spiritual fluid seems to imply that the soul must be regarded as material. Swedenborg had been warned about this in a review of *De Infinito* in a learned German journal in 1735, but he was now inclined to consider the materiality of the soul as more or less a question of terminology: if matter is defined as an extended substance with inertia as an attribute, the soul is clearly immaterial; but on the other hand everything substantial in nature can be said to be matter.[72]

Here Swedenborg alludes to Wolff's distinction between transcendental and physical matter, but he completely disregards Wolff's admonitions not to mix up these different meanings. The reason for this self-assured attitude is that, with the help of the series doctrine, he could distinguish a supreme level of existence, where it is no longer possible to speak of matter in the true sense of the word. This level is life's, and, in its role as the receptive organ for life, the soul cannot be called material, something that also applies in lesser degree to the other mental functions and even to the body:

> For all these live the life of their soul, and the soul lives the life of the spirit of God, who is not matter, but essence; whose esse is life; whose life is wisdom; and whose wisdom consists in beholding and embracing the ends to be promoted by the determinations of matter and the forms of nature. Thus both materiality and immateriality are

> predicable of the soul; and the materialist and immaterialist may each abide in his own opinion.[73]

This conclusion naturally implies a spiritualization of the universe, which is the general prerequisite for the experiences of the seer; and characteristically it is based on a biologically inspired attempt to avoid the ontological difficulties.

The Psychological System

Swedenborg expounded his psychology for the first time in *Oeconomia* and developed it in detail later in several unfinished essays in the 1740s; he also retained it in somewhat simplified form in his theosophical writings.[74] He distinguishes between four different functions of consciousness. Highest is *anima*, the soul, the first and formative substance in the human series. Since *anima* is beyond our actual intellectual life and thus beyond language too, we can have no knowledge of the soul except through analogy or through miracular experiences. Our intellectual process has its seat in reason, *mens intellectualis*, which comprises the second function of consciousness. On closer examination, it also forms a series of coordinated activities, which is particularly clear in the draft of a psychological analysis written by Swedenborg for the *Regnum Animale* series, probably in 1742, and generally known as *De Anima*.[75] The third function he calls *animus*, and this corresponds rather closely with what Aristotle named the vegetative soul: Swedenborg regarded it as the seat of the passions. Finally we have the fourth function, which consists of the sensory organs of the body.

The explanation of this classification is that it permits a parallel between the psyche and the four auras of nature, a combination that earlier was tested in connection with the analysis of the blood and the spiritual fluid. In that way the following scheme could be constructed:

(1) *fluidum spirituosum* or *anima*, the soul, corresponding to the highest aura of nature: its purpose is to represent the universe, to discern ends intuitively, to be the highest form of consciousness and to constitute the determining force of the whole organism;

(2) *mens*, or reason, corresponding to the next highest aura, the magnetic, with the purpose of understanding, thinking, and exercising will power;

(3) *animus*, sensory awareness, corresponding to the ether, its purpose being awareness of sensory perceptions and possession of imagination and desire;

(4) the five sensory organs and their end points in the brain, which correspond to the air (the eye, however, corresponds to the ether). The sensory structure also corresponds to the motor, the last degree of which comprises the muscles.

Naturally, these parallels are not intended to be all-inclusive. When the spiritual fluid is said to correspond to the highest aura of nature and to be counter to the law of gravity, this applies only to its organic aspect, its quality as the highest motor principle of the body, but not to its task of carrying the life of the organism, which presupposes a divine influx. As a psychical phenomenon, *anima* is unaware of its true greatness, its ability to possess all the highest principles of human knowledge, since it is identical with its own behavior. *Anima* can acquire such awareness only by reflecting on the qualities of its subordinates, primarily reason. *Mens* seems from the very outset to be able to follow certain rational principles and to possess logical acuity, which determines its way of perceiving the sensory reports transmitted by *animus*. But this presupposes the presence of the superior capacity that is *anima*.

What Swedenborg achieved here is a kind of compromise between the rationalistic belief in innate ideas in man and the empiricist theory that our consciousness is entirely formed by sensory experiences. This means that, in this great controversy in contemporary philosophy also, he was striving to create a synthesis of opposing views. Aristotle was the father of the empirical thesis to which Locke, the radical philosopher of the day, gave the full support of his authority: "*nihil est in intellectu quod non fuerit in sensibus*," nothing exists in the reason that has not previously existed in the senses. Swedenborg embraced this proposition but only to the extent that the term *intellectus* is confined to a sub-function of the consciousness. Instead, all the principles of human knowledge are innate in the highest mental function; so far Swedenborg adheres more closely to Leibniz and his vision of the soul monad as a *repraesentatio mundi*, a representation of the world. It is significant that when he quotes Locke directly, he does so in order to find support for his own view of *anima*'s function in connection with his description of our intuitive understanding of, for example, mathematical propositions. This is a type of insight that, according to Locke, is possessed by the angels and "the spirit of just men made perfect . . . in a future state. . . ."[76]

The preface to *Oeconomia* I contains a passage of great interest in this connection. Swedenborg describes here two different types of researchers: those who are born with a capacity for observation and a lust for experimentation—those like Malpighi, Leeuwenhoek, Harvey, and Winslow—and those whose strength lies in their ability to unearth the true causes of phenomena by rational analysis of the empirical findings of others.[77] Swedenborg clearly counted himself among the latter category, which is also evidenced by what he wrote about his working methods as a young man. In his description of how this type of researcher experiences a mystical vision of light when, after a prolonged intellectual struggle, he finds a truth—how a mystical radiation penetrates a holy temple in the brain—one is tempted to follow Martin Lamm's example and interpret it as a personal testimony, a reflection of a sign such as resulted from the reading of Leeuwenhoek in Amsterdam.[78] Although nothing contradicts the presumption that Swedenborg in fact was alluding here to Locke's theory of intuition, which could provide a rational explanation of his experiences, it does not mean that Locke was the principal source of his entire psychological thinking. One should perhaps not attribute too great importance to the statements in a rhetorical text like the preface, since they do not necessarily represent more than Swedenborg's characteristic way of describing in mystical and dramatic terms an experience not unusual in a scholar.

A comparison between Swedenborg's presentation and Locke's *Essay concerning Human Understanding* reveals that there is actually no justification for assuming a profound influence. The detailed understanding of the physiology of the brain, which Swedenborg acquired through several years of study, enabled him to maintain a consistently biological perspective, which Locke lacked in spite of his medical training. Swedenborg also stated outright that he found himself in conflict with the prevailing opinion when he placed *anima* above reason, but that he was compelled to do so on psychophysical grounds. Reason grows with the years; we can all observe this in ourselves. The newborn child has no reason at all, but acquires it throughout life by study and experience. The bodily functions proceed normally in the insane, which would not be the case if there were not a soul separate from the sick mind, an *anima* that determines the growth of the organism during the fetal stage when reason does not exist and that functions regardless of the reason being diseased.[79]

It is this implantation of the soul as the developmental principle of the fetus, in accordance with Leibniz's philosophy, which served as Swedenborg's point of departure. Locke's argument should probably be regarded as a decorative support borrowed from the English common-sense philosophy that, as a result of Voltaire's propaganda, was so *en vogue* in the French environment in which *Oeconomia* first saw the light of the day; it is symptomatic that Swedenborg quotes *Essay concerning Human Understanding* in French translation.[80]

If the existence of *anima* can thus be regarded as physiologically confirmed according to Swedenborg, its position in relation to reason means that no being can form any distinct conclusions about the soul as long as he or she is living in the flesh. *Anima* is above the domain of words, since words are tied to ideas, which in turn belong to reason or the senses. We are therefore forced to use scarcely comprehensible words to describe the soul: the soul as a representation of the universe, its intuitive comprehension of ends, its function as the determining factor of the organism, etc. All this could only be explained by the symbols of universal mathematics, and no such symbols exist as yet—one more of the many examples of Swedenborg's dream of another kind of language than the clumsy and inadequate tongue at our disposal.

It was easier for Swedenborg to find distinct and adequate terms for the three lower degrees of consciousness. The task of our reason, *mens rationalis*, is to understand, to think, and to will, and it gets its material, its intellectual ideas *(ideae intellectuales)*, mainly from *animus*, the sphere of sensory consciousness, into which flow the reports from the sensory organs as *ideae materiales*: these seem to be comparable with visual images and appear in finite forms, that is, they have form, size, and position in time and space. As soon as they are transferred to the intellect, however, they lose these delimitations. In cerebro-physiological terms, one might say that *animus* is the body's sensory center in the brain, while *mens* is the soul's activity in the cerebral cortex. Since both these functions of consciousness are situated in the same organ, the boundaries between them will obviously be indistinct.

This becomes increasingly clear as Swedenborg's psychosomatic analysis grew more detailed during the years between the publication of *Oeconomia* and his call to interpret the Bible in 1745, primarily in the posthumously published *De Anima*, where he distinguishes between higher and lower subfunctions in *mens* and *animus*. It is therefore logical that, during his theosophical period, he came to simplify the system by combining *animus* and *mens* in a common function with different levels. This gave him another advantage, namely, the fact that the tripartition of all existing phenomena into natural, spiritual-rational, and divine, which was beginning to emerge in the first draft of the theory of correspondence, could be applied universally.

Before the spiritual world was opened to him, Swedenborg thus worked with four degrees of consciousness, and Augustine occupies the place of honor among the authorities to whom he refers.[81] Swedenborg calls Augustine "*illuminati judicii patrem*," a Father of the Church of enlightened judgment, and this agrees well with what he had learned from contemporary natural scientists, for example, the anatomist Winslow, to whom he so frequently refers and whom he may possibly have met during his long stay in Paris.

These allusions to Augustine are thus of dual interest: they stress Swedenborg's general tendency to seek support from the greatest of classical writers, and they illustrate his affinity with the natural scientists of his day. Furthermore, they recall the complexity of the question of Neo-Platonic influence on Swedenborg. Such an influence was assumed by even the earliest Swedenborg commentators at the end of the eighteenth century, and it gained even greater currency from Lamm's basic research in 1915, not so much in the shape of a direct influence by Plotinus, but primarily through the intermediary of the scholars of the Renaissance and their followers, not least the Cambridge Platonists.[82]

Obviously, there are many important points of agreement, but there is no evidence in Swedenborg's large production of any penetrating studies of a "mystical" natural philosophy of Renaissance–Platonic type. In reality, Swedenborg's reading of Augustine and Leibniz, for example, explains a great deal; but first and foremost we must always bear in mind the research program of 1734. The objective of this study was to prove the immortality of the soul to the senses themselves, that is to say, with factual arguments provided by research in the natural sciences, which would give general validity by way of analogy in cases where language failed.

Swedenborg's philosophical studies, extensive as they may have been especially in the 1740s, were aimed primarily at building an ontological foundation for a systematic investigation of the human body and soul, but he never evidenced any interest in philosophy for its own sake—on the contrary, it is clear that he looked on Descartes, for example, as a natural scientist and never discussed his epistemology. To regard Swedenborg as a link in a *philosophia perennis* with its roots in the esoteric wisdom of antiquity is justifiable to a certain extent, for example, when it comes to his presentation of the doctrine of correspondence, but it should not be thought that Swedenborg was basically interested in theoretical philosophy: he was a pronounced philosopher of nature in the seventeenth-century sense of the term, a searcher for the secrets of nature who, with enormous industry, studied the reports from anatomy classrooms, microscope benches, and observatories.

The psychophysical pattern can be summarized as follows: *Anima*, the soul, is implanted with the father's semen in the mother's egg, which means historically that all human souls are descendants of Adam, who received his soul directly from the Creator, as we know from Genesis. An unfinished essay on the generative organs, intended for the *Regnum Animale* series, reveals that Swedenborg imagined the soul to be the innermost essence of the sperm, formed in the cerebral fountainhead of the spiritual fluid; it is enclosed in a type of corpuscles *(globuli)* in the sperm, the surface of which consists of the spirit of life. The function of the testicles is merely to extract this essence, which is later given its form by the epididymides. At this juncture, Swedenborg categorically asserts that Leeuwenhoek was wrong when he believed he had seen tiny animals or "worms" in the seminal fluid: what he actually saw were formations of these *globuli.*

What is transformed is thus a product of the father's brain, and its first deposit in the egg consists of the first cells of the brain. Although Swedenborg uses the Aristotelian term *punctum saliens* for the first sign of life in the egg, he did not share Aristotle's belief that the heart is the first to enter the egg, but attributed this property to the brain. The transferred soul is the determining life principle of the organism and also possesses its entire store of wisdom from the very outset.[83]

Anima is thus fully informed from the beginning, but *mens,* its first offspring in the organism, is correspondingly ignorant. Its empty pages are filled with the writing of experience throughout life. Through upbringing and living, *mens* proceeds from universal experiences to the particular and individual, and the more richly individualized its conceptions, the better the human being will understand his or her universals. In this way, the child's sensory life comes closer to its soul, which in turn tries to reciprocate, and the meeting place of the two is *mens rationalis,* the intellect that is humankind's privilege. What happens then is that communications are opened between the lower senses of the body and the highest function of its consciousness.

Physiologically, Swedenborg visualized this communication as established with the help of fibers and fluids. All bodily fluids, including the spiritual, flow through fibers of various kinds. This means that they are influenced by the structure of the fibers. Anatomical discoveries have taught us the general properties of the fibers, but we are still dependent on reasoning for the details of their origin. By reasoning we find, in the first place, that the fibers derive from the spiritual fluid itself in its capacity of *vis formatrix* of the organism. Secondly, we can deduce how this transformation from an immensely subtle and volatile fluid into firm tissue takes place. Swedenborg's explanation is that the spiritual fluid is able to undergo endless mutations, that it can expand and contract to adapt itself to every conceivable purpose. But when it has been compressed to the point that it has become a sort of congealed tissue, it loses its distinctive form, which allowed it to live its life in a state of perfection and transfers to a more obscure form of existence.

The smallest fiber that has developed in this way constitutes the last intermediary link between the *a posteriori* sensory reports and the *a priori* orders from the soul. For communications via this link to pass as smoothly as possible, it is essential that there be the greatest possible harmony between the two types of information. In the case of sensory reporting, this means that it should correspond to the nature of the soul, which is order, knowledge, and truth. Consequently, the sensory reports which are best received are those that follow scientific principles—order, knowledge, and truth. When such reports are received, the innermost fibers expand so that they almost regain their original fluid state, though not entirely, since that would break the delicate communication.[84]

In Swedenborg's mind, these scientific principles clearly comprised a sort of classification system, in which individual events are placed in their correct universal categories. The fibers must, therefore, be so formed as to allow for such classification, and this must mean that the child's upbringing is of extreme importance, as already hinted at in *De Infinito* and related draft manuscripts. He does not actually say so, but it is reasonable to suspect that he believed that children should be given an understanding of the theory of series and degrees, which he envisaged as the most perfect classification system, nature's own method of subordination and coordination. A good example of Swedenborg's habit of including trivial observations in important contexts is provided by his interpretation of the young child's desire for knowledge. He claims that if the individual finds learning a pleasurable activitiy, his level of vitality will be raised and the passage of information through the fibers will thereby be facilitated. This is the reason the Creator wanted childhood to be a period of play and *joie de vivre*. In *De Cultu et Amore Dei* (1745), he expresses this thesis in charming idylls that describe and interpret the learning games of the young Adam.[85]

The fiber system that grew out of *fluidum spirituosum* could obviously not be entirely uniform. If it were, the organism would only be able to experience a monotonous *continuum*, since the spiritual fluid is a uniform force in itself. Instead, what is required is a harmonious variation between the fibers, which gives rise to different effects of the same force, according to the conditions of reception. We know from anatomy that the fibers of the brain are softer and more fluid than those of the rest of the body and thus allow for much greater changes in position. The ability to think is also, in Swedenborg's view, located in that part of the brain that is most capable of expansion and contraction, namely, the frontal lobe: in choosing this location, he undoubtedly was inspired by one of his most important anatomical sources, *Neurographia universalis* (1683) by the Frenchman Raymond Vieussens.[86] In this way, the spiritual fluid is able to exert an influence corresponding to the receptivity of the fibers; and it is this form of collaboration between soul and body that constitutes *harmonia constabilita*, the harmony that gradually develops in man.[87]

Clearly the step from this continuous psychophysical flow to its macrocosmic counterpart in the shape of an influx of divine life in the universe was not too great for an ardent reader of anatomical and physiological works, who in addition had been convinced from the outset that existence is a system of analogies, to quote from one of the *Daedalus* essays.[88] One should not without good reason cross the anatomical river to fetch Neo-Platonic water. Furthermore, the term *harmonia constabilita* recalls Leibniz, the second basic influence in this connection. The final chapter of *Oeconomia* includes a central theme that unquestionably originates in Leibniz's theory of harmony.

Swedenborg and harmonia praestabilita

From one point of view, the state of impotence and ignorance in which humankind has found itself since the Fall and from which we strive to deliver ourselves with the help of science is tragic evidence that we were driven out of paradise, that we have left the golden age of innocence and arrived at the hard era of iron—Swedenborg frequently used the classical mythological notion of the ages of the world in his moral reflections. But the lofty isolation of the soul in humankind cannot be explained simply as a result of free choice; the basic cause is ultimately the eternal order which is the manifestation of the Lord himself.

Consequently, our imperfection becomes a link in the Creator's plan, and Swedenborg does not lack for arguments in support of this notion. The original state of innocence, when *anima* communicated directly with the body and humanity thus fully shared its wisdom, was fraught with great danger. In the first place, Swedenborg said he doubted that humankind in that case would have submitted to the conditions of the natural process of birth or of death. If our knowledge were so complete that we could ourselves determine our basic conditions, the effect could have been either that no race would ever have emerged or that the universe would soon have become overpopulated.[89] Today, when conditions of the kind imagined by Swedenborg appear to be within the reach of biological and medical research, his second alternative is well on the way to becoming a reality.

In the second place, language could scarcely have developed, since the exchange of thoughts would have been completely organized from the very beginning and no outer means of communication would have been needed; thus Swedenborg in *Arcana Coelestia* imagined that the first human beings possessed a telepathic gift like the angels. Third, the soul would soon strive to be greater than God himself, and this would be the most horrible of all. Such a sin could never be forgiven, since it would originate in the nucleus of the soul and not from an unenlightened intelligence: there would be no room for mercy. In the fourth place, all bodies and souls would have been equal, which would mean that no societies could have existed either on earth or in eternal life. These and other results of an eternal state of perfection prove that God must choose to allow the perfection of the whole to spring from the variety of the parts and that the purposes of creation can only be realized through an infinitely varied harmony.

All this corresponds closely with the notion of the world expressed by Leibniz in various essays, for example, in the widespread theodicy of the beginning of the eighteenth century: Swedenborg studied this carefully in the Latin translation of 1739. Innumerable individuals all have their given task to make the whole a perfect harmony, even if our imperfect reason cannot apprehend this in the individual case. And how could we do this, when each unit, each monad in Leibniz's terminology, in the last analysis represents one of God's views of his creation? God in his infinite wisdom and perfection regarded this creation from an infinite number of aspects, and each such aspect results in a specific monad. One monad can never be identical with another, according to the axiom that Leibniz called *principium identitatis indiscernibilium*, the law that indistinguishable things are identical. Thereby an indefinite number of individual units develop, which God has predestined to perfect harmony and which reflect his own being.[90] This notion of individuality, which is one of Leibniz's many normative gifts to Romanticism, was embraced by Swedenborg, even though he did not in *Oeconomia* appear to have perceived its full import.

The fall of man also occupies its inevitable position in this harmonious system, and one is reminded of the old liturgical formula "*O felix culpa*" (Oh, happy guilt). It was eternally presaged though not predetermined by God, since God obviously could never positively wish for evil. Sin is the fruit of free will, and this will power was an absolute prerequisite if God were to be able to grant redemption. In this difficult theological question, Swedenborg combined Malebranche and Leibniz, the former in the notion of a redeeming God as the highest of all divine concepts and of the glory of God as the ultimate end of existence, the latter in his vision of a congregation of souls as the manifestation of the glory of the Lord. Life on earth is a seminary to prepare God's people for this congregation. The same symbol can be found in *Oeconomia* in a form which unquestionably derives from Leibniz's monad harmony—no mention need be made of the common inspiration from Augustine's *City of God*—and in the theosophical stage, it was to be carried to extreme lengths in connection with the organism notion, which is one of Leibniz's most significant contributions to the general history of ideas.

Swedenborg returns time and again to the problems of will, but his entire discussion is based on the self-evident assumption that free will actually exists. In his thinking about redemption, a person's actions play a greater part than in orthodox Christianity; and the Lutheran belief in redemption through faith alone *(sole fide)* was repellent to him. This is both a logical result of his early belief in the importance of the brain's being formed in the right way and an indication that his religious upbringing by his father had made its mark: Jesper Swedberg was a stubborn opponent of solifidianism. In general, it may be asserted that in his dynamic system Swedenborg came to attribute greater importance to humanity's active will for redemption than did Leibniz in his more rigid and more optimistic structure of harmony.

On the other hand, this is to a great extent a question of presentation and of points of departure. This is clearly revealed in one section of *De Anima,* Swedenborg's unfinished attempt to produce a counterpart to Wolff's *Psychologia rationalis.* Here, for the first time, Swedenborg applied the notion of correspondence on a large scale to the problem of the connection between soul and body; we have already seen that Leibniz used the French *corréspondance* in a similar context.[91] But by his Latin term *correspondentia,* Swedenborg meant something more technical and concrete. He worked with two types of correspondence, a natural agreement between, for example, musical harmonies and our feeling of pleasure when we listen to them—his early notion of the harmonious relations between the vibrations of strings and the sensory membranes is still valid—and an acquired correspondence, a *correspondentia acquisita,* by which is meant the harmony between, for example, the spoken word and ideas.

The natural correspondence is in explicit accordance with Leibniz's pre-established harmony, while the acquired is the equivalent of Swedenborg's own model *harmonia constabilita.* In *anima,* the soul, there is a pre-established Leibnizian harmony; but between the soul and other functions of consciousness, there exists instead the gradually emerging harmony. Since the harmony in *anima* predates the second, Swedenborg is prepared to speak purely chronologically of a *harmonia praestabilita* in humanity; but this can never be realized as long as we live in our body, for reason and soul never achieve full harmony during mortal life.

However, with the help of his notion of correspondence, Swedenborg was able to reach agreement in principle between his own views and those of Leibniz and Wolff, and he also believed that the most profound meanings of the other hypotheses of psychosomatic interaction, *influxus physicus* and *causae occasionales*, could be embraced in the same notion. This would provide a way out of the deadlock, which Swedenborg regarded as primarily a battle of words. His argument is a fine example of his struggle for synthesis; the fact that at this stage it resulted in a model closely related to the Leibniz–Wolff intellectual structure is important and irrefutable.[92]

The Doctrine of Forms

The theory of correspondence is the most important of the doctrines that Swedenborg believed would enable him to carry out the research program of 1734: to prove the immortality of the soul to the senses themselves, to establish a psychological science based on natural science. But it is not the only one, and Swedenborg's theory of form, his *doctrina formarum*, deserves comment before we examine in detail the origin and meaning of the theory of correspondence. It was first developed in the treatise on the fibers of the body, which was intended as transaction III in the *Oeconomia* series but was never completed.[93]

The point of departure was one of Swedenborg's many attempts to combine anatomy and ontology, and the motive was his desire to determine the form of motion that characterizes the simple fibers of the body. The anatomical evidence derives from Leeuwenhoek, whose views on the form of the blood vessels and the nerve fibers, circular or spiral as in a shell, Swedenborg found especially interesting. Since not even the strongest microscope could visualize the simple fiber, Swedenborg was compelled to arrive by deduction at a ladder of forms, and he compares his method with the progression of infinitesimal calculus from power to power.

The lowest form is the angular, which is bound by straight lines with no common center. It is found in nature in the primitive polygons, which are the principles of salts and which influence our sense of taste and smell. The next form is the circular or spherical, which is the principal form of material motion.

The third form is the spiral, or, to be consistent in terminology, the perpetuo-circular. For an element of infinity is added here in comparison with the immediately preceding form, which in this case means that the focus of these forms constitutes a spherical center corresponding to the position of the earth in the atmosphere.

This third stage of form comprises the active forces, and we have already reached so high that words or figures can no longer provide an adequate description. Obviously, it becomes even more impossible when we arrive at the perpetuo-spiral, the form of vortical motion. In this connection, Swedenborg explicitly expresses his acceptance of Descartes's theory of vortices, even though he admits that it has been abandoned by several modern authors. He regarded this vortical form as the higher form of the active forces, that is, the form of the *conatus* within the active forces; he thereby tried to accommodate the notion of *conatus*, so important in Wolff's works, to the scale of forms, and we shall soon find it again in the various presentations of the correspondence theory.

As in the theory of series and degrees, Swedenborg stresses that each form contains a *repraesentatio*—this is Wolff's terminology; and, true to his synthetical ambition, Swedenborg also refers to *exemplar*, the term of the ancients of all the lower forms, hence a reflection or image of the succeeding ones. The higher one comes on the scale, the more rich in meaning the forms will be. Immediately above the vortical form, we find the perpetuo-vortical, also called the celestial form. This is the natural principle of the active forces and is the highest of all nature's forms. Swedenborg regarded it as identical with what Plato in the *Timaeus* called "The One" and in the *Parmenides* "The First," and with Leibniz's unit or "monad" and Wolff's "simple substance." This also means, however, that it lacks shape, extension, and weight: at this level, we have shed matter and reached the point at which heaven flows into nature.

But the ladder is still not complete. At the top, there is the perpetuo-celestial or spiritual form, which too can be perpetuated to the supreme stage, the Divine. The spiritual form can obviously not be described in words, but it seems that, in Swedenborg's imagination, it also represented a broad spectrum from the highest abstraction down to the borderline of the form immediately below it. Its primary function is to communicate the Divine to human souls, a task it has in common with the Word of the Lord.[94]

In his description of the Jacob's ladder of forms, Swedenborg worked from physics, but there are also corresponding forms of vital character, biological and psychological. The construction was carried out analytically, but Swedenborg also attempted to observe his form scale from the synthetical angle—he often uses the terms *analysis* and *synthesis* in the same sense as Newton, to differentiate between inductive and deductive reasoning, and he constantly stresses that the analytical approach is the only one open to humankind since the Fall.

From the synthetical perspective, the form scale appears to emanate from the Divine, but this does not mean that the Creator is responsible for the existence of imperfections and deficiencies at the lower levels of the scale. What has happened is that the nature of an originally perfect form has undergone a change. This is a fate that can only befall the vital forms, which have been accorded freedom of action, namely, angels and human minds; but it may produce secondary effects, a transformation of the elements of nature—the Apostle Paul teaches us that the whole creation has been groaning in travail together (Rom. 8:22).

Such essential transformations are tragic consequences of human evil, but they are by no means unique. In addition to them, there are normal accidental changes, that is, metamorphoses that do not affect the inner essence of the forms but only their superficial, temporary qualities: the scholastic influence is notable here as in many other instances when Swedenborg carries out ontological analyses, and this, of course, relates to his reading of pre-Cartesian handbooks and of Wolff.

The power of accidental changes is an integral part of the perfection of the higher forms and enables them to create new substances that are given the closest lower form. Swedenborg uses an example at an elementary level: a particle of steam consists, in part, of a combination of air and ether masses, in which the ether, normally vortical, takes on the spiral form as long as it moves within the steam; as soon as the combination is dissolved, the constituent parts return to their original forms. The highest forms are thus always present in the innermost conceivable part of the innumerable combinations, and all that truly exists in them is the highest form of nature, the celestial or, to use his own terminology from the theory of series and degrees, the first substance. In order that all lower forms will not disintegrate, a fixed order is required in which each and every thing occupies a given position.[95]

Swedenborg's *doctrina formarum* can be traced to his own speculations on corpuscles and particles and to the evidence from ancient and modern metaphysics which he himself cited, but its foundation is obviously the same as for his entire system, his certainty of universal order. What is interesting, however, is not this certainty as such—it was shared by the majority of his contemporaries—but its individual formulation. Swedenborg's ambition to reach a synthesis of old and new is striking here as in the psychophysical interaction discussions, but almost equally apparent is the inspiration from what Swedenborg himself in a moment of exasperation likened to obscure oracular pronouncements, the Leibniz–Wolff system.[96] The scale of forms is Swedenborg's version of the crystalline universe of the monads, their hierarchy from the petrified life of the stone to God's all-inclusive consciousness.

5

Universal Philosophy and the Theory of Correspondence

The Background

When Swedenborg in 1734 defined the objective of his future researches, to prove the immortality of the soul to the senses themselves, he soon discovered that he needed an insight into all the existing natural sciences, primarily the biological disciplines. He was prepared to acquire these insights regardless of the effort it cost him, but this was not the basic problem. In his determined struggle to penetrate to the heart of the unknown, one deficiency revealed itself to be increasingly crucial, namely, the lack of an adequate means of expression, literally the instruments with which to capture the innermost secrets of the life of the soul.

Bearing in mind Swedenborg's mentality, we can easily understand that he would first search for these instruments in the realm of mathematics. The first time he defined the conditions for successful psychological research, he began by speaking of a new philosophy. Even if we could succeed in mastering all sciences, he wrote in the paper of 1738 usually called *The Way to a Knowledge of the Soul*, this would still not be enough, unless we could manage to achieve such a high level of understanding that all branches of science could be combined in a universal science, with the help of which all particulars could be restored to their universals.

This imaginary science he called *philosophia mathematica universalium*, a mathematical philosophy of universal concepts. Without its help, we might as well try to reach the moon as to dissect the soul, since it constitutes the very analysis that makes it possible for the soul (*anima*) to comprehend its own nature.[97] The theory of series and degrees represents an attempt to formulate this philosophy. As was established in chapter 4, Swedenborg regarded this doctrine as a representation of universal order, a means of assigning phenomena to their right categories; and it thus constituted a prerequisite for a logical and natural means of expression.

In his desire to construct a language as exact as that of mathematics, Swedenborg adhered on the whole to the central ideas of the masters of the seventeenth century. The hope of overcoming the dreary consequences of the Tower of Babel through logical-mathematical artificial languages is, of course, older than the seventeenth century. It can be traced back to the linguistic logic of Scholasticism, particularly Raimon Lull's *Ars magna* from the end of the thirteenth century, but it waxed stronger with the great mathematical advances of the 1600s. Descartes, one of the greatest mathematicians of the century, believed in principle in the possibilities of creating a universal mathematical language. The newly founded Royal Society engaged in various efforts to increase the precision of language, the most interesting of which is John Wilkins' great work *A Real Character and Philosophical Language* (1668). This study was the most advanced attempt ever made to achieve both a logical system of classification and an ideographic writing related to the interpretation of Egyptian hieroglyphs and Chinese characters at the time.[98]

Wilkins's artificial philosophical language proved to be no more than a new approach and a challenge. Swedenborg probably knew of it, but here too Leibniz was his most important source of inspiration—when Helvétius mentions these efforts to create "*une langue philosophique*" in *De l'esprit* (1758), he refers only to Leibniz, which is characteristic of the time.[99] Leibniz was early fascinated by the idea of the perfect logical language, and he made several attempts to establish its laws. In this connection, he obviously proceeded from the same basic concepts as in his monadology, namely, that the predicate is contained in the subject, or, as expressed in terms of the monad theory, the monad is a *repraesentatio mundi*, a representation of the universe.

To Leibniz, definition became the principal task for all science; and, if it is true that every subject contains within itself all its possible predicates, definition is an attempt to make these predicates manifest. In addition to nominal definitions, which unequivocally differentiate the defined from other things, Leibniz worked with real definitions, which describe the construction of the phenomenon defined in such a way that it can be concretized: geometrical definitions belong to this category, as do chemical formulas. These real definitions are a part of what Leibniz called *ars inveniendi*, the art of invention, and play a central role in the so-called *ars combinatoria*, the art of coordinating concepts.

Thereby we have come closer to the formalization of the language that Leibniz tried to achieve with his *characteristica universalis*, the specific system of characters that would permit calculation with concepts in analogy with algebra. The system of characters is a prerequisite for the art of combination, which is dependent on the monad theory in substance. According to it, the world is a vast system of independent entities, which change their relationships to one another but remain unchanged in their essence; what is new qualitatively develops when quantitatively comprehensible elements form new combinations. The aim of philosophic analysis is to disclose the simplest basic notions, which shall not be purely intellectual constructions but shall be abstracted from a concrete experience of reality, to the extent possible: most of our thinking is concerned with concepts, which cannot be given concrete form or be established with material norms, for example, God, spirit, intelligence, will, substance, force. This means that man's search for truth is, in the last analysis, a matter of the correct arrangement of concepts, and Leibniz obviously had in mind an analogy with mathematics.[100]

The foregoing is a summary of Leibniz's logical philosophy, but it might just as well have originated in Swedenborg's discussion of the theory of series and degrees, especially if one lays greater stress on the concrete concepts, the "images" of things. Leibniz's project of a "conceptual algebra" assumes that simple concepts can be expressed by numerical symbols and thus composite concepts can be regarded as additions of the simple ones. But he soon discovered that such simple operations were not adequate; it was essential, for example, that the order of priority between the concepts themselves be determined. He never doubted, however, that it would be theoretically possible to invent a mathematics of concepts, and he even believed that this had been partly achieved in the Chinese system of characters—Leibniz was one of the numerous Sinophiles of his day and entertained an exaggerated belief in the wisdom of the Chinese.

Naturally he never succeeded in achieving his dream of *characteristica universalis*, and modern critics have had no difficulty in proving that there exists only a handful of branches of science in which an extension of the conceptual apparatus from a few simple basic concepts is possible: geometry and mechanics were Leibniz's own models.[101] What is essential and what was decisive for Swedenborg, however, is that Leibniz worked so intensively to construct a method with which one could imitate the monad's quality of a *repraesentatio mundi*, and a language that unequivocally and completely would represent the psychic laws as mathematics represent those of nature. There is no doubt that the *philosophia mathematica universalium*, which would make possible the soul's self-analysis—the soul that is *repraesentatio mundi* for Swedenborg—is a direct descendant of the Leibnizian dream.

It is not difficult to detect the relationship, but what about the detailed genealogical connections? To begin with, we must recall that Swedenborg was fascinated by mathematics at an early stage, the science for which he confessed "an immoderate desire" in a letter of 1711. Soon after his return from his long stay abroad in the 1710s, he produced a number of small mathematical papers, one or two articles in

Daedalus Hyperboreus and an algebra textbook (*Regel-Konsten Författad I Tijo Böcker*, Uppsala, 1718).

Experts have pointed out that Swedenborg was well-informed in contemporary mathematics and that his knowledge of the subject was probably superior to that of his countrymen at the beginning of the eighteenth century. At the same time, even a layman can see that the textbook contains so many arithmetical errors that it is completely useless for its purpose. Even more serious, he seems to have put far too great stress on analogies.[102] On the other hand, it is probably this very tendency to regard nature as "one complex of analogies" that prepared the way for his later attempts to construct an all-inclusive system of mathematics. Swedenborg learned from his great masters Descartes and Leibniz the elements of analytical geometry and differential and integral calculus, the greatest discoveries of seventeenth-century mathematics that seemed to have revealed intoxicating approaches to infinity and to new worlds of symbols.

In his studies of the physiology of the brain and of psychology, Swedenborg was constantly forced to work with infinitely small and ephemeral materials, and he was indeed soon drawn to the brink of infinity. It is scarcely surprising, therefore, that one often finds references to infinitesimal calculus in his writings between 1734 and 1745, and it is equally natural for a disciple of Descartes to refer to analytical geometry. In the first cerebro-physiological work, probably written in 1738, he speaks of a mathematical universal philosophy or a *philosophia graduum*, which will lead us to ever more profoundly hidden phenomena and with whose help we shall be able to see things and their series in almost the same way as geometry can detect what is hidden in a figure through algebraic analysis.

In another passage in the same text, he states that it is essential to distinguish the degrees of organic nature and their mutual order just as we regard the infinite as higher than the finite or differentials as higher than integrals.[103] In the final psychological chapter of *Oeconomia*, the theory of series and degrees is considered as a vital premise for all science, but it cannot in its own terminology express anything above the level of ordinary things. To reach higher, we must have a *philosophia universalium mathematica*, which can not only express loftier ideas in a simple manner with letters which progress in order, but also restore them to a kind of philosophical calculus reminiscent of infinitesimal calculus in form and rules.[104]

There are many statements of this kind, most of them equally difficult to understand, but during the same period we also encounter critical objections to what he regarded as inadequate mathematical analogies on the part of his illustrious predecessors. The most striking example is found in a partly extant tract on the harmony between soul and body written in the beginning of the 1740s, in which Swedenborg severely criticizes the *harmonia praestabilita* model.[105] To this criticism he adds an assumption that Leibniz and Wolff had found decisive inspiration in differential and integral calculus. Arguments based on that sort of calculus and on pure analysis are inapplicable to real entities, he claimed on this occasion. Criticism of this kind coming from Swedenborg sounds like an exceptionally flagrant example of inconsistency, but what he wanted to say was probably a variation of his chronic objection to scholars who construct theories based purely on rational considerations without making sure that they correspond with the evidence of experience. In the introductory chapter of *Principia*, he stresses that geometrical and philosophical arguments must rest on a solid empirical foundation, and to these basic principles he considered himself to be faithful.[106]

The Search for philosophia mathematica universalium

It is obvious that Swedenborg was aiming at a mathematical language from the very outset, but there are also ways of determining the immediate sources of his obscure words about a *philosophia mathematica universalium*. The three last propositions in the final chapter of *Oeconomia* I, which bears the Wolffian title "Introduction to Rational Psychology," are of great interest in this connection. They assert the following: (1) to the same extent that it ascends through degrees, nature mounts from the sphere of the particular and generally used expressions to that of the universal and supreme, until, finally, in the highest region of the body where the soul resides, there is no mortal language (*loquela corporea*) to describe the essence of the soul, even less things that are still more elevated; (2) therefore, one should evolve a *philosophia universalium mathematica* to express what words cannot through *notas characteristicas et literas*, numerals and characters, which resemble in form the algebraic infinitesimal calculus; (3) such a philosophy, correctly interpreted, could become the science of sciences. The last two propositions are supported by a quotation from Wolff's *Ontologia*, which in the original contains the following passage: "Since mathematical knowledge of things consists of knowledge of their quantity, and qualities have quantity, i.e., degrees, which are made comprehensible with the help of numerals, it is clear that mathematical knowledge of qualities is possible."[107]

Wolff's proposition is clearly a direct portent of Swedenborg's intentions with his theory of nature's series and degrees, but Swedenborg's quotation is—as quite frequently—inaccurate. In his commentary, Wolff begins by giving a few examples to show that this qualitative mathematical knowledge was already established in certain cases and mentions some of his own many textbooks as sources for such phenomena as the intensity of light, the gravitation of solid bodies in fluids, the weight and elasticity of air, all of them measurable. The problem, according to Wolff, is to discover dimensions for qualities in general, as well as laws governing variations of the phenomena, which can be described by curves. This presupposes an established and unequivocal understanding of the qualities.

All this is summarized in the following reflections, which Swedenborg quoted in part:

> Among the desiderata is a science which would deliver the general principles of the knowledge of finite things; a science from which the geometricians might draw their measures, when desirous usefully to exercise their calculations in the mathematical knowledge of nature, exactly as they turn to Euclid's *Elementa* to find the principles of calculus. And this science would have a better title to the name of *universal mathematics* than a science of quantities in general, or of

> indefinite numbers, since it would deliver the first principles of the mathematical knowledge of all things. . . . Ah, would they put their minds to such a task, they who in their search for glory in literary matters are convinced that nothing remains for them than to denigrate the efforts of others and blacken their reputations!"[108]

It is not clear how Wolff envisaged this coveted discipline in detail, but his aim was clearly to construct certain universal principles for the measurement of qualities, whereby they could be defined in comprehensive terms, along the lines of the correspondence between air pressure and the length of the column of mercury in a barometer. The goal of Wolff's desideratum is nevertheless as distinct as it is lofty: the achievement of the true mathematical principles of psychology and the philosophy of nature; or, expressed in another way, Newton's program could be applied to the psychical universe and be expanded within the physical world to reveal the innermost causes of existence.

Bearing in mind what we know of Swedenborg's scientific ambitions ever since the 1710s, it is not surprising that he diverted his search for literary laurels in the direction of universal science that Wolff had pointed out to him. There is no doubt in my mind that here we have the essential text among his sources, but it must be complemented by others of Wolff's many works to which Swedenborg so frequently referred during the busy and restless years up to 1745. Since it may be questionable whether Swedenborg actually studied Leibniz's universal language project in the original, it is important to note that it is reproduced in considerable detail in the psychological work by Wolff from which Swedenborg was taking extensive notes in 1734.[109]

There the disciple defines the master's concept of *ars characteristica* as the science that explains the use of symbols to indicate things or the sensations they evoke. This science was still mainly in the developmental stage, according to Wolff; but, when completed, it would embrace the universal principles of mathematics and grammar, both working with symbols. Through this art, humankind would acquire symbols for qualities, just as algebra offers symbols for quantities. This is, of course, much more difficult and presupposes—to repeat what was said in *Ontologia*—that we first learn how qualities can develop from other qualities. We understand the process with regard to quantities, and this is described in algebra and logic by combinations of simple symbols.

In that function the imagined universal science is known as *ars characteristica combinatoria.* Algebra is actually a part of the science that, according to a quotation from Leibniz by Wolff, would be able to distinguish rational ideas from both sensory perceptions and fantasies—precisely the function that Swedenborg assigns to his psychological concept of *intellectus purus*, the pure intellect, as we shall soon see. With the help of the art of combination, we could also, according to Wolff, trace the causes of things to their source in the invisible world, and we could unearth hidden truths through a kind of algebraic calculation. But before this can be done, we must establish certain indissoluble basic concepts in the same way as mathematics did in the numerical unit of arithmetic—still another reminder of Swedenborg's terminology in *Oeconomia*, when he tried to apply the term *unitas*, unit, to the simple substances of the series. Finally, this science would enable us to transform our present symbolical knowledge into intuitive or direct perception of the truth.

Swedenborg almost certainly read this summary at the very significant point in time when he was beginning to prepare for his biological and psychological researches. In all likelihood, it was Wolff's summary that caused him to try to construct the Wolffian desideratum. This effort can be found in a small manuscript, probably finished in the beginning of 1740, immediately after the completion of *Oeconomia* I.[110] The most remarkable feature of this fragment is its title, *Philosophia universalium characteristica et mathematica*, which reveals that the author was trying to achieve an unequivocal system of symbols without having to invent new words. The symbols would have to represent concepts with absolute clarity, and Swedenborg tried the easiest approach, namely, by using the initial letters of the Latin words in question.

The choice of points of departure is characteristic of Swedenborg's physiological bent: he applied the designation *S* to the purest blood (*sanguis*), *SS* to intermediary blood, *SSS* to red blood, and *SSSS* to clotted blood, and set up corresponding sequences of signs for arteries (*A*), muscles (*M*), and nerves (*N*). At the highest level, *N* and *A* are identical, namely, the simple fiber that, according to an earlier work, is also a product of *S*; but the determinative processes take separate tracks and thus give rise to different series calling for special designations. The goal appears to have been to create symbols for the basic concepts of the organism that could indicate their correspondence at different levels of significance. Incidentally, it is in this fragment that Swedenborg used the word *correspondentia* for one of the first times. He also added a few designations for quantity and connections within the various series.

The brief fragment is difficult to understand, partly because it was broken off before Swedenborg arrived at any arithmetical operations which might indicate how he planned to continue. Obviously, it is a kind of combination art, in which the coordination of symbols reflect the fundamentals of Swedenborg's interpretation of the organism and its origin, and so far it is clear that he wanted to apply what he had learned from Wolff to a limited area. He certainly was aware that these problems were still far from solution. In the last chapter of *Oeconomia* II, he reverts to his universal philosophy, "which shall be enabled not only to signify higher ideas by letters proceeding in simple order, but also to reduce them to a certain philosophical calculus, in its form and in some of its rules not unlike the analysis of infinites; for in higher ideas, much more in the highest, things occur too ineffable to be represented by common ideas."[111]

Here, however, there is a difference in nuance compared with earlier statements. In the first place, Swedenborg complains about the difficulty of the task, which is quite understandable in view of his recent experiences; second, he refers not only to Wolff but also to Locke who, in his *Essay concerning Human Understanding*, included a brief section on what he called *semiotiké*, the doctrine of signs. This reference may indicate a trend away from conceptual algebra *in abstractu* and toward speculations concerning languages already in existence, thus a step in the direction of his future exegetic activity. If that was the case, it is natural that this trend made its appearance in a psychological context in which mathematical analogies seem particularly inappropriate.

The last noteworthy discussion of *philosophia universalium mathematica* is also presented in a psychological context, namely, in the unfinished tract *De Anima.* Here Swedenborg develops an idea hinted at in *Oeconomia* II: that this mathematical philosophy of universal concepts is identical with the language of the angels—this is actually a logical result of the reference to Locke's theory of intuition just mentioned. In *De Anima,* Swedenborg tried to make a detailed analysis of the psychic functions, divided into the principal groups defined in the psychological chapter of *Oeconomia*: *anima,* the soul; *mens,* the mind; *animus,* the lower mind; and the sensory organs of the body. Each of these includes subfunctions—in *mens,* for example, imagination, thought, and pure intellect.

It should be noted that now, after the intensive studies that one can follow in the excerpts, Swedenborg had arrived at certain shifts in the boundaries between the levels of consciousness in relation to the preliminary determinations in *Oeconomia. Imaginatio,* imagination, is an inner and higher sense of sight, a simultaneous reproduction of memories of a succession of sensory reports. *Cogitatio,* thinking, forms new ideas from these recollections in such a way as to reactivate the mathematical analogies: the recollections are compared with figures in a calculus, from which the mind creates equations through a kind of analysis related to infinitesimal calculus.[112] These equations are treated at a higher level as simple ideas, just as entire equations in algebra can be handled as units. In Swedenborg's imagination, types of parentheses are apparently formed around the mental sequences; and this is undoubtedly both a characteristic and suggestive way of expressing the same conviction that, in his theosophical writings, caused him to attribute innumerable inner meanings to every word, indeed every letter in the Bible.

The more complicated these equations, that is, the more ideas that form higher units, the closer we come to the sphere of pure intellect. Physiologically, this occurs as changes in the fibers and membranes of the cerebral cortex; we recall that Swedenborg had come close to the modern view of cells as the smallest components of the brain, and in themselves reflecting the whole brain they constitute a microcosmic correspondence to the entire brain. Each one of these infinitesimal cerebral parts contains as its supreme organ what Swedenborg calls the simplest cortical substance which is the organ of the pure intellect, molded in the highest form of nature or, according to the doctrine of forms, the celestial. He tried to give a concrete image of these subtle materials through the angel's dissection in *De Cultu et Amore Dei* mentioned earlier.[113] This simple cortical substance is so high on the scale that it cannot be named with terms intended for lower substances, and Swedenborg, therefore, gave it a special designation, *intellectorium.*

Pure intellect has thus been interposed between *anima* and *mens* according to the original design, and one of its functions is to have access to the science of sciences, that is, *mathesis universalium.*[114] The associations with Wolff's *Psychologia empirica* already described become even more confirmed by using the term *intellectus purus*, which in this context represents a mental function so abstract and subtle that it is only attracted by pure mathematics; but Swedenborg may also have been inspired by Malebranche's use of the term.[115] The pure intellect contains the principles of science from birth and thereby determines our entire experience of the world, but it does not possess either the purest intelligence or the immediate connection with the Divine, both of which qualities are still reserved for *anima*, the soul. Obviously, Swedenborg felt a need to introduce a connecting link between the soul and the mind which would lessen the distance between the two functions: to use his own mathematical analogies, one might say that he wanted to work with more precise limits than previously.

In this way, he was also able to accommodate the simple cortical substance of the smallest cerebral parts in his psychological scheme. Not the least important aspect of this observation is that the reader is reminded of the fundamental premise for Swedenborg's psychology and later for his experience of the spiritual world, namely, that he was constantly haunted by more or less precise ideas of the behavior of the physiological foundation, the fibers and fluids of the brain. A note in the *Diarium Spirituale* (Spiritual Diary) in August 1748 tells how he had seen the thoughts and speech of the spirits gyrating in a manner highly reminiscent of the convolutions that exist in the human brain but that can never be understood by the mind, since the form of the brain corresponds with that of the spiritual world.[116] This note exposes Swedenborg's special dilemma: the brain and the psyche cannot be thoroughly analyzed without an understanding of the spiritual world, but the form of the spiritual world is clearly incomprehensible without some knowledge of the physiology of the brain.

In Swedenborg's eyes, which saw the spiritual fluid as identical with the soul without implying any materialism or subjective idealism, all ideas comprise changes of condition in the brain. Hence, since such changes can be understood only through a description of their forms, it is theoretically possible to capture the conceptions of the mind in a calculus, i.e., a system of universal mathematics is conceivable. Thus far, Swedenborg maintains his old position in *De Anima*, and he even writes that he has made certain experiments in that direction. We have already seen how inadequately they met the requirements for microbiological precision and conceptual clarity.

But having said that, he introduces a new element. The tremendous problems involved in the work no longer seem to be worth overcoming, since the results that could be anticipated are uncertain, and the most insignificant error in calculation could lead to innumerable mistakes. For this reason he declared, "I forbear making the attempt, and in place thereof I have desired to set forth a *Key to Natural and Spiritual Arcana by way of Correspondences and Representations* which more quickly and surely leads us into hidden truths. Since this doctrine has hitherto been unknown to the world, it behooves me further to dwell on it."[117] With these words, Swedenborg intimated the doctrine that made his name best known, particularly in literary contexts, namely, the theory of correspondence. Bearing in mind how this doctrine was used by symbolist poets to justify poetry's aspirations to express truths that cannot be formulated by science, we find it indeed ironic that the original theory of correspondences to such a great extent emanated *aus dem Geist der Mathematik*, to travesty their contemporary Nietzsche.

Before we study the basic text of the doctrine, it is only fair to alleviate the irony with other passages from Swedenborg. In the pure intellect in *De Anima*, Swedenborg had found a temple courtyard, where the speech of the angels resounded in the algebra of the universal philosophy of which he dreamed, but how could this seraphic tongue be made intelligible to the human mind? Previously, Swedenborg relied entirely on the artificial mathematical language of the future, but now he suggests other possibilities: it can be done through images such as we see in our dreams, through parables, and through fables like those of the period immediately after the golden age; and it will then be the task of the mind to interpret these representations, as the ancients interpreted the obscure sayings of the oracle, communicated by the Pythian priestess from the Delphic tripod.[118] Thereby Swedenborg himself indicated that poetry can convey knowledge inaccessible through other media. The relationship between pure mathematics and the highest form of poetry in Swedenborg's theory of correspondence is one of the elements of the visionary's interpretation of life that brings us to our own century, and the doctrine deserves more detailed discussion.

The Concept of Correspondence

If we are to establish the import of Swedenborg's concept of correspondence, the obvious source is the treatise in which he first discussed it in a relatively thorough manner. This was done in an unfinished essay, probably written in 1741 or 1742, entitled *A Hieroglyphic Key to Natural and Spiritual Arcana by way of Representations and Correspondences* (*Clavis Hieroglyphica* to quote the initial Latin words of the title). Swedenborg never published this essay, and it was not printed until 1784 in London in a highly inaccurate Latin version.[119]

The title may sound obscure, but it reflects rather adequately the extremely complex background that Swedenborg was trying to summarize and formalize. The phrase "hieroglyphic key" arouses associations with the antique tradition of mysticism and magic, related to the interpretation of ancient Egyptian writings, and thereby also to the special combination of poetry and graphic art that flourished during the Renaissance and the Baroque era, namely, emblematics. The term "natural and spiritual arcana" indicates the juxtaposition of nature and spirit and is the hallmark of Swedenborg's scientific approach; the word *arcana*, secrets, alludes to his lofty ambitions. The concept "representations" belongs to rationalistic psychology and immediately recalls the basic monadologic notion of the monad as a *repraesentatio mundi*; from there it is a short step to Leibniz's speculations concerning an artificial language of which we have already heard.[120]

The mathematical associations come very quickly to those who read this little essay with unjaundiced eyes, while the literary ones are almost nonexistent: readers are probably at a loss to understand how so many poetical trees could sprout in such an arid desert. The fragment consists of twenty-one examples, in which propositions from three areas are compared and discussed. The domains are the natural, the spiritual-intellectual, and the divine; and the author tries to demonstrate that, by exchanging the central terms, the validity of these propositions can be transferred to their correspondences.

The sixth example can serve as a prototype. The first proposition, which belongs to the sphere of nature, reads in translation: "From effects and phenomena, a judgment is made concerning the world and nature, and from the world and nature, a conclusion is made as to effects and phenomena." This proposition, which is a description of the inductive and the deductive method or, in Swedenborg's Newtonian terminology, analytical and synthetical, is then contrasted with the second one from the spiritual sphere: "From actions and inclinations, a judgment is made concerning man and the rational mind; and from man and his mind, when known, a conclusion is made as to actions and inclinations." Finally, we have the third proposition, which belongs to the spiritual life: "From works and the testimonies of love, a judgment is made concerning God, and from God, a conclusion is made as to his works and the testimonies of his love."[121]

In this way, Swedenborg establishes propositions of identical formal structure in which a number of words are exchanged with their counterparts at other levels. Here he follows a principle to which he had adhered on many other occasions, namely, that one cannot use the same expressions, the same words, when one moves from the inanimate to the animate or from the sphere of the senses to that of the mind. In this very example, he drew an analogical—or, in his words, harmonious—conclusion from the three propositions, which he expressed as follows: "As the world stands in respect to man, so stand natural effects in respect to rational actions. As man stands in respect to God, so stand human actions in respect to divine works." In his comments on his line of reasoning, he suggests that we call the world *M* (*mundus*), effects *E* (*effectus*), humankind *H* (*homo*), and actions *A* (*actiones*), whereby we can formulate the analogy as *M:H=E:A*. He states categorically that this constitutes the first foundation of *mathesis universalis* and promises to demonstrate later how these terms can be combined with others and multiplied and create an analytical equation.[122]

This demonstration never materialized, but there is no doubt that Swedenborg was referring to what Wolff called *ars characteristica combinatoria*; all the same criteria are present. He took a small step forward by pointing out that still another analogy can be drawn from the preceding ones: "As the world is to man so man is to God." Man is thus the proportional mean, which is a mathematical way of expressing a belief that God's relationship with nature can only exist through man and that the perfection of nature depends on man's and vice versa. Later we shall encounter more stimulating expressions of the same thought, which is one of Swedenborg's fundamental propositions—one may ask whether anyone else ever carried the idea of humanity further than he did in his completely humanized spiritual world.

In the same way, the other examples in *Clavis Hieroglyphica* formulate central correspondences between a conceptual apparatus of natural philosophy and the terminology of the higher levels, but they have no new observations to offer. Instead, the purpose is to construct a method that, in Swedenborg's eyes, will unassailably demonstrate the relationship between the different degrees of existence, a deductive process that reveals universal order with mathematical exactitude. The essential, of course, is to produce the interchangeable key words, the true linguistic correspondences; and the author does append such a list to his manuscript. This is probably what he hinted at when he published for the first time his ideas of correspondence in a more specific sense; the examples cited hitherto all derive from posthumous publications.

This occurs in a very interesting and, in part, surprising context, namely, in a chapter on the human kidney in *Regnum Animale* (The Animal Kingdom), vol. I (1744). After long physiological discussions based on reports from his scientific authorities, Swedenborg stresses that humankind undergoes constant regenerative processes in body and soul, in the same way as the blood is cleaned and rejuvenated during its circulation. Thus, the spiritual life is always represented in the corporeal, which is the import of the words concerning searching of "the reins and the hearts" in the Revelation of St. John (2:23).

These words were written by Swedenborg's most exalted predecessor, who described what had been revealed to him in visions, in the same manner as the Lord was to command Swedenborg to do in a very near future. It is not unusual for him to seek support from the Scriptures in his scientific works; on the contrary, this happened quite frequently and is in line with the research program outlined in 1734. However, the situation was quite different at this juncture, which is particularly evident from the commentary; being a crucial text, it deserves to be reproduced at length:

> In our Doctrine of Representations and Correspondences, we shall treat of both these symbolical and typical representations, and of the astonishing things which occur, I will not say in the living body only, but throughout nature, and which correspond so entirely to supreme and spiritual things, that one would swear that the physical world was purely symbolical of the spiritual world: insomuch that if we choose to express any natural truth in physical and definite vocal terms, and to convert these terms only into the

> orresponding spiritual terms, we shall by this means elicit a spiritual truth or theological dogma, in place of the physical truth or precept; although no mortal would have predicted that anything of the kind could possibly arise by bare literal transposition; inasmuch as the one precept, considered separately from the other, appears to have absolutely no relation to it. I intend hereafter to communicate a number of examples of such correspondences, together with a vocabulary containing the terms of spiritual things, as well as of the physical things for which they are to be substituted. This symbolism pervades the living body; and I have chosen simply to indicate it here, for the purpose of pointing out the spiritual meaning of *searching the reins.*[123]

This is still another example of how Swedenborg promises great things, sends up trial rockets for his readers from rather shaky launching pads; as far as we can judge, he had not yet had time to complete more of the preliminary research than is found in *Clavis Hieroglyphica* and in the relatively extensive grouping of biblical texts under various correspondence headings found in the collection of excerpts, cod. 36–110, to which we shall return later.

What is important, however, is that he now promised a correspondence dictionary, which would seem to indicate that he had abandoned the idea of trying to construct a universal mathematical system of symbols. When he wrote his draft of this system, one of the arguments in favor of the selection of symbols was that there would be no need to write great dictionaries, even though at that time newly created words would be involved, which obviously would make the project even more difficult, as Wilkins's universal linguistic plan clearly illustrates. *Regnum Animale* also contains remarkably few traces of the necessity for a *philosophia mathematica universalium* stressed so frequently in earlier works. Instead, the correspondence theory now really worked as the alternative conceived by the author in *De Anima*: transpositions of letters of the ordinary alphabet would be adequate, which means that the way had been cleared for a biblical interpretation that necessarily proceeds from actual texts.

Hieroglyphs and Emblems

The title *Clavis Hieroglyphica* contains a reference to the symbolical writing of the Egyptians, and this is explained in Example XVI, the annotations of which discuss spirits and angels; the choice of topic was in itself not strange or offensive at the time, but the argumentation is characteristic of Swedenborg. These beings, unlike human ones, were created with their minds fully developed, and since they are spirits, they are superior to nature. Nevertheless, they know everything about natural things, which must mean that there is a correspondence and harmony between spiritual and natural things and that, consequently, there is nothing in nature that is not an image of a spiritual prototype, an idea. It should be stressed that this theory of spirits was considerably revised during the theosophical stage with regard both to the creation of spirits and to their comprehension of nature, which gives the system a completely different logical consistency than in this outline, which most closely resembles a kind of vague Platonism.

The goal of Swedenborg's reasoning, however, is to arrive at the ancient Egyptian wisdom which, he believed, embraced a similar theory of correspondence: these *correspondentiae* were designated with "hieroglyphic characters" of different kinds, which could express not only natural phenomena but at the same time spiritual things. His reference for this Egyptian science is "a whole book by Aristotle," by which he meant a Neo-Platonic treatise wrongly attributed to the philosopher, *De secretiore parte divinae sapientiae secundum Aegyptios.*[124] That the work was not by Aristotle he discovered later, and it can scarcely have given him any major impulses, even if the title undeniably has a Swedenborgian sound. At the same time, references to the writings of the Egyptians recur in the entire theosophical production in connection with Swedenborg's description of the historical background of his hermeneutics. On one occasion in the first and biggest exegetic work, *Arcana Coelestia* (1749–1756), he bewails the fact that the understanding of the innermost meaning of the Scriptures had been lost in Europe: "It had yet been present in Chaldea, in Assyria, in

Egypt and in Arabia, and from there in Greece; in their books, emblems and hieroglyphs there still exist such things."[125]

Swedenborg makes a great many similar references to Oriental and Greek mythology, emblematics, and hieroglyphics, in his efforts to prove that his exegesis in fact revives a primitive comprehension, which the churches dissipated during the course of their tarnished history. Apparently, he wanted to affiliate himself with an esoteric tradition, but the references are so general that it is risky to try to determine the sources. Ever since classical times, scholars have written of this sort of submerged wisdom, a primitive revelation that continues to survive beneath the changing surface of faith and knowledge. Nor is the specifically linguistic form in which Swedenborg expressed the notion original. It is thought-provoking, though, to encounter it in the middle of the eighteenth century, as it is a personal reflection of the new interest in humanity's primeval language, an interest associated with such names as Vico, Rousseau, and Herder, and that was to become of great significance to literary Romanticism.

The secret of the hieroglyphs was not definitively unveiled until 1822 by Champollion, but innumerable efforts to do so had been made ever since late antiquity. The mysterious and tantalizing aspect of early Egyptian writing derives mainly from the fact that, in addition to its phonetic function, it also serves an ideographic purpose: each character represents a concrete material object and also a sound that reproduces the sound in the prototype that gave it its name (not vowels, however, which remained undesignated as in Hebrew). The phonetic purpose of this writing was obscure, and interpreters of both ancient and Renaissance times concentrated on the ideographic aspect, which, in addition, they observed through allegorically ground lenses: the Egyptian world of mythology was understood in the same symbolical and allegorical categories as the Greco-Roman.[126]

When the scholars of Europe in the fifteenth century learned of the only fully extant ancient hieroglyphic treatise, Horapollo's handbook *Hieroglyphica*, the result was not only a revival of the allegorical hieroglyphic discussion, but also impulses for a new literary genre, emblematics. In the original version in Andreas Alciati's *Emblemata*

(1531), the emblems were composed of three elements, a brief fable or allegory in Latin verse, a corresponding symbolic image, and finally a short comprehensive motto. It was thus a matter of combining words and images to form a concrete representation of an idea, usually a moral proposition. Behind this lay a number of ideas related to Egypt: the image was a counterpart of the hieroglyphs and the text corresponded with obscure proverbs, known as Pythagorean maxims and presumed to contain the most profound wisdom stemming from Egypt; it was an old and widespread belief that both Plato and Pythagoras acquired their wisdom during long years of study in Egypt.[127]

In later developmental stages, the image might be left out and replaced by other literary devices in which analogies, similes, and metaphors were of central importance; and the concept that the purpose of poetry is to give symbolical shape to supreme wisdom—to be a *poesis parabolica* in the terminology of the day—walked hand in hand with these tendencies. But it was not only poets and painters who made use of the special modes of expression provided by emblematics, and innumerable examples of hieroglyphic symbolism can also be found in works on theology and the philosophy of nature. Before the invention of alphabets, humanity knew God through hieroglyphics; and what are the heavens, the earth, and all beings but hieroglyphics and emblems of his glory, in the words of an English emblematicist in 1635.[128] In the same year, the associations of natural phenomena with hieroglyphics are carried out consistently in Sir Thomas Browne's basic contribution to the history of ideas, *Religio Medici* (The Physician's Religion); when as modern an author as Baudelaire coins the phrase "*tout est hiéroglyphique*," all is hieroglyphic, in his important essay on Victor Hugo (1861), he expresses the same nature-symbolism idea.

Emblematics was also put to use in pedagogics at a very early stage, primarily by the Jesuits; they used images and mottos in their teaching to arouse and maintain the children's interest, and one of their most significant followers in the educational dialogue of the seventeenth century was Amos Comenius, theologian and educator of the sect of the Bohemian Brethren, who during one period was employed by the Swedish government, a man to whom we shall return later. It is thus clear that hieroglyphic symbolism was useful in a number of different and also conflicting ways. Mario Praz has emphasized that a contradiction characterizes the entire purpose of the art of emblems. On the one hand, its proponents want to create an esoteric language reserved for the few, a poetry for a few initiates, a learned and speculative art that can capture what the blunt tools of science cannot grasp. On the other hand, they want to make moral and religious propositions intelligible to all, including children and the illiterate mob; in other words, they adhere to the tradition of the medieval *Biblia pauperum*, the Bible of the poor, or of the stained glass and sculpture of the cathedrals.[129] These contradictory tendencies recur constantly, interwoven with each other. Practically all books on emblematics refer to Horace's *utile dulci* program (that is, combining profit and pleasure) to justify the method of making moral messages intelligible and digestible, but at the same time emphasize the noble origins of emblematics in hieroglyphic antiquity and interpret its symbolic language as the tongue of God and the angels.

The background just briefly summarized was certainly familiar to Swedenborg. Among his excerpts, he noted the title of the standard work on emblematics, Picinelli's *Mundus symbolicus*, in a Latin edition published in 1695; he frequently encountered hieroglyphics and emblems in that part of his reading which we can verify.[130] He also made exciting use of these terms in several of his works. In one passage of *Arcana Coelestia*, he states that symbols like those used in coronations were usually called emblems because nothing was known of correspondences and representations; obviously, he regarded his *doctrina correspondentiarum* as the true idea on which all these forms of symbolic language were based.[131]

But the bonds can be tied even more firmly and more exact sources established. The name Amos Comenius was certainly well-known to Swedenborg, and he presumably was familiar with Comenius' great textbook *Orbis sensualism pictus* (The World in Pictures) and the pansophical dreams behind it and other comparable works.

Comenius belonged to the fascinating category of seventeenth-century scholars who strove to construct a scientific system that would combine all scientific disciplines in a universal knowledge, a pansophy. With his doctrine of series and degrees and in his universal mathematics, Swedenborg himself was a descendant of the same brotherhood.

In any case, Comenius became a reality to him when, in 1734, he excerpted Wolff's *Psychologia empirica* in Leipzig. Among the quotations, we find several notes on the hieroglyph concept, which are exceptionally interesting and directly applicable to *Clavis Hieroglyphica*: they yield many clues to the mystery. When Wolff analyzes the human imagination, he also gets into what he calls hieroglyphic signs; by this he means signs through which a thing can be attributed a significance beyond itself. As an example he takes a triangle, which is used to designate the Holy Trinity. Swedenborg noted one of Wolff's definitions: "If a phantasm is so composed that by the similitude which its constituent parts bear to the intrinsic determinations of some given thing, the latter can be inferred from the former, the phantasm has a hieroglyphic signification and is composed by force of the principle of sufficient reason."

From the annotations on this and certain subsequent definitions, Swedenborg observed that "the ancients represented dogmas and historical matters by hieroglyphic figures. This was a familiar practice with the Egyptians, and some say the same thing of the Chinese. Comenius exhibits the human soul hieroglyphically."[132] This reference leads directly to *Orbis pictus*, in which Comenius tries to illustrate the concept of soul by allowing a number of points to create a human form—in this way, he also gives shape to the indivisibility of the soul (the points), its substantiality, its union with the body, and the quality bestowed on it by Scholasticism of being both in the whole body and in its every part (the silhouette of the body).

All phantasms are naturally not so highly complex, and Wolff stresses that there are various degrees of perfection among them. The highest perfection results in a total correspondence between the details of the phantasm and the thing that it concerns, and then we have what Wolff calls a perfect hieroglyph. If this represents the concepts with which a concept is defined, it can replace the definition; such hieroglyphs thus convey knowledge. According to Wolff, this was practiced by the Egyptians, but, unfortunately, we can no longer fully grasp their methods. He further explains that the letters of the Hebraic language have sometimes been believed to possess hieroglyphic significance, and he refers to a work by one of his own teachers, the well-known theologian and statistician Caspar Neumann's *Clavis Domus Heber* (The Key to Eber's House). Finally, Wolff emphasizes that we must differentiate between the truth that is inherent in the thing described and that which lies in the hieroglyphic sign. He writes that he himself assumes that images of deities originally were hieroglyphics and designated divine qualities, but that this hieroglyphic import was eventually forgotten so that we now are astonished that human beings could envision such absurd gods.[133]

This, of course, is a clear parallel with Swedenborg's views of the original manifestation of correspondences, which later was lost and misrepresented in various forms of idolatry until he himself was called to revive the truth. There is no doubt that it is the hieroglyph concept in this sense that explains the choice of words in the title *Clavis Hieroglyphica.* There is nothing to suggest that Swedenborg was interested in the Egyptian written symbols as such: instead, he concentrated exclusively on the insights into nature as a reflection of transcendental life from which they were believed to have sprung. His reading of Wolff recalled the literary and pedagogic hieroglyphic tradition of the Renaissance, and it is significant that he encountered this reminder in a psychological work by the contemporary scholar whom he most admired at the time.

The Categories of Correspondences

With this we return to Swedenborg's own hieroglyphic key, which, despite its fragmentary and mathematically abstract formulation, is just as ambitious and stirring to the imagination as any of the writings of the emblematists. Each one of these twenty-one examples sets up correspondence relations, which led to formulation of a number of rules for the interchange of key words in the propositions. But Swedenborg would not have been the man he was if he had not tried to create an even firmer system from his own observations. At the same time, this attempt reveals that, in his preliminary studies, he had already invaded more portentous domains than the natural-philosophical and psychological ones in which the examples originated. For the tract results in his setting up a number of categories of correspondences, the majority of which reach beyond the examples.

There are four categories. The first is called *correspondentia harmonica,* and it resembles most closely the type that was exemplified in *Clavis.* The relationship between it and the mathematical background is clear, but is absent in the remaining three. By harmonious correspondence is meant the variant that prevails between light, intelligence, and wisdom; between modifications in matter, sensations, and thoughts; between visual images, ideas, and rational arguments—terms that can be compared with those in a mathematical analogy. The attribute *harmonica* marks the association with the psychophysical thinking in *Oeconomia,* where the model of interaction, *harmonia constabilita,* was presented as the true solution of the problem of the intercourse between body and soul.

The three other categories of correspondence, however, lead us in somewhat different directions. Instead of mathematics, they are suggestive of ancient hermeneutic principles and also of symbolistic theories of the nature of poetry. Together they seem to comprise the theoretical premise for the translation of the tongue of the angels, which Swedenborg in *De Anima* assumes can be made in our minds and will offer interpretive principles for dreams, parables, and ancient myths.

The second category is thus known as *correspondentia allegorica.* In the preliminary stage, it was given the attribute *parabolica,* the same term the Renaissance art philosophers used to characterize the kind of poetry that possessed the mysterious profundity of the hieroglyphs.[134] In *Clavis,* however, its use is confined to the biblical allegories; in the volume of excerpts, most of the parabolic material of the Gospels was put under the same heading.

The third category is called *correspondentia typica*, and this attribute points directly at biblical interpretation. What Swedenborg was trying to achieve is usually known as typological or figural exegesis, even though he extended the meaning beyond what is customary, that is, an exposition which in the Old Testament identifies direct precursors of Christ and the New Covenant: when Abraham was prepared to sacrifice his only son Isaac as a burnt offering (Genesis 22), this was a *typus* or *figura* of what would happen at Calvary, even in such detail that the son had to carry the wood to his own pyre. To this Judaic hermeneutics, which of course has been of extraordinary significance in the history of poetry—we need only mention Dante—Swedenborg added tales from the New Testament that prophesied the future kingdom of God and the angelic society.[135]

The fourth and last group in *Clavis*, *correspondentia fabulosa*, is associated directly with poetry. In this category, Swedenborg unites myth and poetry with man's dream world, as occurs in the pure intellect according to *De Anima*; his high esteem for ancient mythology and poetry as conveyers of knowledge harks back to his earliest literary experiences: paraphrases of Ovid can be found in some of his juvenile Latin poems.[136]

Two significant propositions are laid down following these different definitions of the categories of correspondences. In the first, he declares that we have the right to assume that the universe is filled with *typi*, presages, even if we are unaware of them in practice. The present moment always includes the future, and as long as the vital fluid from divine providence flows out into the world, everything consists of connecting links (*contingentia).* In both his way of thinking and of expressing his thoughts, Swedenborg once again adheres to Wolff.[137] The second proposition states quite briefly that it is permissible to interpret the Word in this way. Here the exposition is abruptly terminated, and all that follows is a relatively summary list of corresponding concepts.

The second proposition is obviously a later addendum, although it has not been observed by Swedenborg scholars; and we cannot know when Swedenborg received permission to interpret the Bible according to his correspondence system, nor how he experienced it.[138] It might have been a very late appendix, made toward the end of his life, when he promised to give an interpretation of Egyptian hieroglyphics with the help of the doctrine of correspondence, a promise that he never fulfilled.[139] However, he can scarcely have been referring to his first attempts to group biblical texts in this pattern in the volume of manuscripts cod. 36–110, since they must have been produced before *Clavis.* There are several reasons for drawing this conclusion. For example, the name of the second correspondence category was changed from the original *parabolica* to *allegorica*; in this manuscript he also worked with two further groups of correspondences and representations.

As already mentioned, there is a pronounced difference between the first correspondence category and the other three, but they are all regarded as applicable to the Scriptures. They were, in fact, utilized for this in cod. 36–110; and even though it is sometimes difficult to understand why a biblical passage is set in one or the other group, a pattern and a possibility of explaining the origin of the system are discernible. The manuscript cod. 36–110, which is highly rewarding as source material and which is readily available in English translation, consists primarily of excerpts from psychological and metaphysical texts. A list on one of the first pages indicates the sources to be consulted: the classics are represented by Plato, Aristotle, and Augustine, while the modern names comprise Grotius, Descartes, Malebranche, Leibniz, Wolff, and two lesser lights, the German Leibnizian Bilfinger and the Swedish philosopher Andreas Rydelius. To these Swedenborg adds the Holy Scriptures, which is not surprising in view of his earlier declarations of intention.[140] A great deal of space is devoted to the question of the interaction between soul and body. In this highly controversial set of problems, the notion of harmony is of central importance to the scholars cited and to Swedenborg himself, but there he may also have encountered the otherwise rare Latin term *correspondentia.* Following his long quotations from source materials, Swedenborg jotted down in an empty space still more notes, together with his own reflections, under a heading that is a kind of collection of material and a preliminary demarcation: *Typus, Repraesentatio, Harmonia, Correspondentia.* His own speculations draw parallels between the capacity of the soul to represent everything that occurs in the body and God's relationship to the universe, in the same way as Malebranche did in his *Recherche de la vérité* (1674): God's conceptions of the world can not be separated from his essence, and all creation therefore exists in him; our conceptions of the creation are only limited conceptions of the Creator.[141]

Later in the manuscript, when Swedenborg pursues the lines of thought suggested in the heading and in the reflections on Malebranche, this takes the form of a complete survey of the Bible from one aspect, stated at the outset: that it is essential to learn to understand the import of spiritual language in order to avoid fatal misunderstandings and religious disputes. This systematization of the Scriptures begins with the category *correspondentia harmonica*, after which the others follow at pre-established intervals. The first heading is thus closely related to the psychological excerpts; under it, Swedenborg reports on texts that combine natural phenomena and psychological concepts: light, day, night, water, air, blood, body, spirit, endeavor, will, love, etc. For these he seeks spiritual counterparts, which can give the texts a meaning acceptable to him.

The second and third categories he took from traditional exegesis, while the fourth, *correspondentia fabulosa*, mainly consists of stories of dreams and visions in the Bible. To these he adds episodes that he explicitly interprets as parallels to the ancient fables, including the stories of the creation of man, the serpent in paradise, and the Tower of Babel.[142] This interpretation is of special interest as an explanation of his daring to create his own mythic version of the birth of the first human being from an egg carried by the Tree of Life in *De Cultu et Amore Dei* a few years later.

The most likely interpretation of this first systematization of correspondences seems to be the following: In the course of his detailed and extensive excerpt work on metaphysical and psychological problems, designed as a preliminary to the scientific works that would complete *Oeconomia* and, ultimately, the superhuman research program of 1734, Swedenborg made a thorough study of the Bible from a number of different aspects. At the same time, he became familiar with various attempts to formulate the interaction between soul and body with the help, among other things, of the concepts *harmonia* and *correspondentia*; and he learned from Descartes's pupil Malebranche that the psychophysical interrelation is parallel to God's relations to his creation. While Swedenborg was engaged in this work, the idea came to him to try to systematize God's own Word with the aid of the same concepts and other biblical categories of symbols, with which a man of Swedenborg's clerical background obviously was familiar.

That the harmony category comes first underlines the correlation between the correspondence theory and the philosophy of universal concepts; through its analogous character it comprises "the first foundations of *mathesis universalium*." The combination of scientific theorizing and biblical exegesis is not sensational in itself, but represents a step in the same direction followed by a number of Swedenborg's most illustrious colleagues in the previous generation: Steno, Swammerdam, Newton, Boyle, Hooke, all of whom devoted much energy to interpreting the Bible and, to that end, "took the trouble of learning the holy tongue," as Boyle expressed it.[143] Swedenborg, it is true, had studied Hebrew as a young man, but this first systematic review of the Bible was made in Latin translation, and several years were to pass before he began his study of the original texts—bearing in mind what he had learned from Wolff about the hieroglyphic significance of Hebrew, one can perhaps relate his study of the language to the sphere of *philosophia mathematica universalium.*

However, there was an interval of several decisive years between the tentative speculations in the volume of excerpts and *Clavis Hieroglyphica* and the conviction of a divine call in the exegetical works. The most interesting traces of these years can be found in the *Journal of Dreams*, and their literary fruits are discernible in one of Swedenborg's most original and beautiful works: the fragmentary treatise on the creation and the first human beings, *De Cultu et Amore Dei* (*The Worship and Love of God*), Thus, the years 1743–1745 deserve special consideration.

6

The Religious Crisis

Journal of Dreams

The minutes of the meeting of the Royal Academy of Sciences in Stockholm on July 2, 1743, noted that the members listened to a short summary of the work "which Mr. Swedenborg had compiled on Anatomy and now was disposed to make available to the public in print," whereupon the academy, of which Swedenborg had been elected member in 1741 on the proposal of Carl Linnaeus, had expressed its gratification. Shortly thereafter, Swedenborg departed for The Hague, where he was to publish the completed parts of *Regnum Animale*, the work in question.

In the first pages of his travel diary, Swedenborg wrote briefly of the people he had met on the journey and of what he had seen of towns and countryside. He arrived in Hamburg on August 12, 1743; during his five days there, he spent his time in the company of several Swedish travelers of the highest social standing—his birth and profession gave him access to these circles. The most flattering incident was his audience with the newly elected Crown Prince of Sweden, Adolf Frederick, who graciously permitted him to tell of his plans for publication and to show reviews of earlier works.[144] The experience, duly transformed, was relived in several of the dreams in the following year.

These introductory notes give the same picture of Swedenborg's social environment and scientific aspirations as the earlier travel diaries, but they came to an end after a few pages. In their place came a series of almost illegible notes on various dreams, which Swedenborg recalled from earlier years, after which there is a kind of summary of his mental state. Swedenborg was puzzled about several things: his desire for recognition as a scientist had disappeared since he had arrived at The Hague, as had his "taste for women . . . which has always been my chief passion."[145]

These are categorical statements, but they were subsequently contradicted in dreams from March to October 1744. In reality, his ambition is revealed as his most decisive moral problem, as a terrible hindrance to what he craved, filiation to God. If it is true that the fifty-six-year-old man no longer worshiped Venus in deeds, which is questionable in itself, his subconscious had nevertheless retained the most concrete traces of that "chief passion." In the "wakeful ecstacies" of which he speaks in addition to the normal dreams and which should possibly be regarded as a kind of trance, psychologists of religion have detected a phenomenon common in mystics, one which usually contains an element of sensuality. Martin Lamm has rightly stressed that Swedenborg was not only fully aware of this connection between divine and mortal love, but also described it without circumlocution.[146] A review of the contents of the *Journal of Dreams* will reveal that both the dream work and the interpretations were to a great extent determined by his scientific experiences, which as we have seen comprised studies of procreation.

The first dated dream, during the night of March 24, 1744, is characteristic. Swedenborg thinks he is standing by a machine, which is operated by a wheel, whose arms grip him and lift him up so that he cannot free himself. This is a mechanical nightmare, natural to a dreamer trained in engineering, to be added to the many examples of more conventional anxiety dreams, for example, the experience of falling into abysses and pits. Following the dream, Swedenborg suggests its meaning, as he regularly does in these notes: it either meant "that I ought to be kept longer in straits" or it referred to what he had just written about the lungs of the fetus in the uterus—this alternative recalls that his anatomical point of view was basically mechanistic. He finally decided that the dream could be interpreted in both ways.[147] Similar dreams, immediately associated with his scientific work, recur through the *Journal of Dreams* and comprise approximately one-third of the 150 dreams mentioned.

The work on different sections of *Regnum Animale* went on during the entire dream crisis, but the intensity and attitude vary. When Swedenborg began to keep a detailed record of his dreams in March 1744, he was in The Hague and was preparing the second part of the great work. The dreams that recurred practically every night reveal frantic worry concerning the completion of the work. Most striking, however, is the struggle to achieve "faith without reasoning," a goal particularly difficult for intellectuals but also paradoxically attractive for them. This agony reached a climax at Easter 1744 with the first great vision of Christ: Swedenborg was hurled to the floor in his chamber, felt his joined hands embraced by another's hand, and at last looked into the face of Christ, who was smiling warmly.[148]

But not even this incomprehensible grace gave the dreamer peace. He did not question the experience itself, but he was uncertain about the spirits that had acted on him: were they evil forces that wanted to bring about his downfall by making him believe that he had been chosen by God? The pitfall of spiritual arrogance plagued him. In addition to his constantly recurring scientific ambition, he now also was exposed to the lure of the halo and of martyrdom, and he was scarcely helped by his decision not to make public what his nights had revealed to him. The journal remained unknown until 1859, when the royal librarian in Stockholm bought it from an estate and published it in a small edition.

Nor did his passionate prayers to the crucified Jesus Christ give Swedenborg the peace he craved, except for brief moments. They too carried temptation; and his father, the bishop, scolded him in a subsequent dream as one reprimands a little boy: "You are so excitable, Emanuel." When Swedenborg analyzed this dream, he deciphered it as a warning against the cult of the cross and abandoned the idea of hanging a crucifix in his room.[149] That Swedenborg was marked for all time by his Protestant origin is quite obvious.

In mid-April Swedenborg wrote that his crisis had lasted for twenty-one days; the same day he asked his friend Joakim Preis, the Swedish Minister to The Hague, to arrange for him to take communion once again. The next night Swedenborg had a hideous nightmare. He saw an executioner roasting decapitated heads and throwing them into a bottomless stove. This horrible scene is even more ghastly because the executioner was a big woman, who never stopped smiling and who had a little girl at her side; Swedenborg writes that he was "at intervals in interior anxiety, and at times in a state of despair." Attendance at a church service the day after renewed his distress at not being able to silence his mind's arguments against faith.

But at last, out of his fear, sprang the longed-for "faith without reasoning." *Pura fides*, the pure faith in which the child has unquestioning confidence, cannot be influenced by either positive or negative thoughts. The achievement of this state is not helped by theological arguments of the kind that had certainly troubled Swedenborg in endless Protestant sermons, nor by attempts to prove correspondences between Christianity and philosophy: this notation is the first example of Swedenborg's strongly negative view of the so-called natural theology, so popular at the time. What man needs is to abandon his "adoration of his own intelligence." But no one can do this by him- or herself, particularly not the learned; it is an act of grace by God.[150]

What Swedenborg had now come to believe corresponds with the dream experiences, but not even at this moment of capitulation did his scientific training cease to make itself felt in the shape of reservations: "Much of what I have experienced agrees with this, perhaps also the roasting of so many heads, which were the food of the Evil One, and their being thrown into a stove."[151] One might think that Swedenborg should by now have reached the stage where research no longer had any appeal; what remained to be done for one who had gained "faith without reasoning" except to preach this faith? What purpose would still be served by his attempts "to prove the immortality of the soul to the senses themselves," as his future research program in 1734 was formulated, when the whole truth is contained in the pure faith of the child? And there are hints in this direction in the *Journal of Dreams.* Only a few days after he had won his insight, he considered stopping his work and returning home; nothing came out of this, however. The work on *Regnum Animale* was continued for a time, possibly at a faster pace, and Swedenborg appears to have convinced himself that this was his true vocation in the service of the Lord.[152]

At the beginning of May 1744, Swedenborg moved to London, where he visited the church of the Moravian Brethren in Fetter Lane. As has been rightly emphasized by the latest Swedish editor of the *Journal of Dreams,* Lars Bergquist, Swedenborg had clearly been tempted for a time to become a member of this congregation, and it is conceivable that he actually applied for admission. However, he did not join them—or maybe was not accepted by them—and his later *Spiritual Diary* contains numerous critical attacks on the Moravians. The obscure notes on the Brethren in the *Journal of Dreams* nevertheless give an indication of how Swedenborg was seeking other ways of realizing the Christian life, and there are signs that he had considered taking orders; a conversation with his father in a dream earlier in the spring can scarcely be interpreted otherwise.[153]

The struggle against temptation and the lures of the devil continued through the remaining months of the *Journal of Dreams* period, but it did not have the same anguished character as before. Swedenborg's inner joy was so strong, particularly when he was alone, that he compared it with heavenly happiness on earth. It could only be threatened by relapses into the pursuit of worldly fame, which occurred on several occasions: he was especially pleased with something he had just written, he had boasted to someone about the work he was about to produce, he had listened to the deliberations of the medical faculty and hoped to hear himself cited as the foremost among anatomists.[154]

But the dreams always brought him back to the right track. He was writing the third part of *Regnum Animale* with great confidence. In the middle of June, he dreamed of a beautiful grove of fig trees and of a great and handsome palace overlooking the orchard. He interpreted the palace as the plan of his work, a work directed at the spiritual, which was symbolized by the grove of fig trees. The same palace recurred in a dream on the last day of September; Swedenborg considered this dream to be a sign that what he had just written with God's help on the organic forms would permit him to achieve even more glorious vistas.[155]

This information is of the greatest interest. The section *De formis organicis in genere* (On the organic forms in general), constitutes a comprehensive application of the early outline of the doctrine of form to the human body, and it means that Swedenborg had revived the theory of series and degrees, the foundation of universal philosophy and the theory of correspondence. Its general premise is the teleological nature of organisms. The reasoning starts by a proposition directly based on the doctrine of series in *Oeconomia*: the organic forms of the body are perfect in relation to the degree of simplicity of the forms that make up its units. The lung's unit is the air receptor; the muscles', the motor fiber; the brain's, the brain particle and the simple fiber, etc.

Swedenborg recalls the mathematical inspiration of this organic hierarchy by mentioning how arithmetical units and numerals are applied in mechanics, astronomy, anatomy, and other sciences. During the determinative process, the units are normally raised to the third power. Here again we find the same tripartition on which the analysis of the blood was made in *Oeconomia* and which occurs in the draft of the *philosophia mathematica universalium* in 1740. The various series of the units, like the entire organism, are united by special ties and links. The result is the most perfect inner coherence of the units in all the series, and these, in turn, form a corresponding pattern—the figures are different, but they remain in the same relationship to one another.[156]

This summary of the organic forms is Swedenborg's last attempt to capture the laws of the biological microcosm in a mathematical model; and it is significant that, after his overwhelming spiritual experiences, he still feels his perspectives expanding "to even more glorious things" when he allows his anatomical knowledge to be arranged in categories of figures and chains of analogies. This is strong evidence of the force of his youthful vision of everything as a composite analogy, and it is difficult to overestimate the significance of the mathematical inspiration behind the correspondence theory in particular and also of the entire conception of the world of spirits.

As reflected in the *Journal of Dreams*, however, this was the last time that Swedenborg felt any satisfaction with his work on *Regnum Animale.* His mathematical enthusiasm had disappeared a few days later, and from that time on, he regarded his pride in his writing as a dangerous temptation: in the dream it is represented by a trek across thin ice toward a gaping hole.[157] Instead, he turned to making plans for an entirely different book, a *liber divinus*, which would deal with God's love and the right way to worship him. But Swedenborg still felt doubt. One night he dreamed that he was lost in the fog, another time that he had been admitted to the kingdom of innocence; he was torn between completing his original plan and becoming involved in the new project, which sometimes seemed like a "toy" in comparison with the old one.[158]

But by the last notation from the night of October 26, 1744, all doubts had been overcome. Swedenborg now believed that he had received divine guidance in his new work, which would contain nothing "of the articles manufactured by others," that is, it would not be based on scientific source materials as the preceding studies, but, under the guidance of Christ, would be founded on what he himself thought and experienced. The decision was confirmed in a dream recalling a childhood memory of a market held in his father's house in Uppsala—Swedenborg was probably about ten years of age—but also by a more obscure phenomenon, which is frequently mentioned in the literature. When Swedenborg awoke and saw the light of day, he had a fainting fit or a "*deliquium*" and threw himself prostrate on the floor. He recognized the symptoms from a similar attack when he began *Oeconomia* in Amsterdam, and he had his interpretation immediately at hand. His head was to be cleansed and purged of anything that could prevent his thoughts from penetrating the subject to be treated.[159]

Unfortunately, this marks the end of the *Journal of Dreams*, and we therefore no longer have the same possibility of studying the progress of the work on *De Cultu et Amore Dei.* The first sections were published in the spring of the following year, as was the third volume of *Regnum Animale*; but both these works, which came into being under the impression of the dreams and visions of Swedenborg's crisis, remained fragmentary. We shall soon revert to *De Cultu*, but first a few comments on the *Journal of Dreams.*

Swedenborg's mental status

Since its publication almost a century and a half ago, the *Journal of Dreams*, a difficult and obscure notebook, together with the extensive notes on his communion with the spirits during the theosophical period in *Diarium Spirituale*, has been the main source for psychological and psychiatric research on Swedenborg. There is no doubt that the *Journal of Dreams* reflects a profound psychical crisis, but it only allows us to observe the culmination in any detail. We thus know little of the origins, even though certain elements in the list of dreams are traceable far into the past, in some cases to the years around 1720. As an old man, Swedenborg explained that he had communed with the angels even as a child and that his parents had been amazed at his gifts.[160] Jesper Swedberg, however, did not mention anything of this in his memoirs, which makes it difficult to put any great faith in the anecdote. It seems rather to be an understandable adjustment of the facts to fit into a hagiographic tradition, an attempt to view his entire life as an uninterrupted process of illumination.

But the fainting fit in Amsterdam is one landmark, even if there is some question as to the actual date on which it occurred. As is often the case, Swedenborg is somewhat vague, and, judging by him, 1738 is the earliest possible date. But he was not in Amsterdam that year, though he did visit the city on other occasions around that time, and he returned there during the latter part of 1739, when he drew up plans for *Oeconomia Regni Animalis.* In all probability he meant the last-mentioned sojourn, and this signifies that the experience was related to the "sign of truth" he said he possessed in his summary of the corpuscular philosophy at the beginning of 1740.[161]

How, then, did he experience this *signum*? He noted in *Diarium Spirituale* in 1748 that, long before the spiritual world was opened to him, he had experienced a marvelous light when he wrote certain sections of his scientific treatises; and we have already noted similar claims in the preface to *Oeconomia*, even though they are highly inexact. It appears to have been a matter of some variety of photism, and Swedenborg apparently gave the episodes an increasingly concrete definition in connection with the belief held by him and by many—perhaps most—of his contemporaries in the existence of spiritual beings. Based on Swedenborg's accounts of his fainting fits and his respiratory manipulations, certain psychiatric researchers have assumed that a mild form of epilepsy was an ingredient in his illness. Indeed, most specialists tend to agree that Swedenborg was mentally disturbed after 1744, even though the diagnoses vary.

It is obviously impossible for the layman to take a definitive stand on the question of Swedenborg's state of mental health, but certain reflections may be justified. Since he was not examined when alive, the diagnoses are based entirely on impressions gained from his writings and on more or less unreliable statements by witnesses. The latter include a number of sensational claims, for example, that Swedenborg had an acute attack of insanity in London in 1744, that he suffered from delusions of persecution, hurled himself to the ground in the filthy streets, appeared naked in public, gave his money away to the poor, etc. It is quite conceivable that the psychical crisis manifested itself in violent forms on one occasion or another, but these sources are hardly reliable.

Furthermore, an occasional violent episode can scarcely reveal much about the person's mental state during almost three decades as a visionary. Friends and foes alike have affirmed that Swedenborg behaved as behooved a learned elderly gentleman and that he caused no offense except through his stories of the spiritual world, which he was occasionally pleased to relate in society. In addition, he performed his duties as the head of his family in the Swedish House of Nobles, and his contributions on economic matters were not only taken seriously but were regarded by many of his contemporaries as highly qualified.

Emil Kleen, a Swedish psychiatrist who most energetically has supported the theory of Swedenborg's insanity, has also stressed that his diagnosis, *paraphrenia systematica*, which according to Kraepelin's terminology is a paranoiac psychosis with hallucinations, is not necessarily characterized by recognizable symptoms of mental illness. Too, Kleen has had to admit that the psychosis did not prevent Swedenborg from developing his religious views until he was well into his eighties.[162] This fact is obviously an essential one and also embarrassing to the advocates of the insanity theory; and it is easily verified by the works that survived him, even though they are difficult to interpret from the psychiatric point of view.

As Martin Lamm has pointed out, the claims of many eighteenth-century as well as modern writers that Swedenborg was mentally disturbed are based on much-too-broad definitions: all of Swedenborg's religious fantasies, all his deviations from the lukewarm normal were regarded, *a priori*, as proof of insanity, and the term is thus deprived of any significance.[163] The most problematic documents, of course, are his spiritual diaries and his use of these notes in the *memorabilia* recounting experiences in the spiritual world in his theosophical works. The unused sections include sexual and excremental fantasies about people identified by name, which occasionally recall a writer of the eighteenth century who in other respects was his absolute antithesis, Marquis de Sade.[164]

In these hideous descriptions of the hellish existence of relatives and historical personages, Kleen and many others saw all-too-clear manifestations of paranoid obsessions, and this is certainly an understandable attitude. On the other hand, in evaluating Swedenborg's total personality, one should not overemphasize things that he was generally wise enough to conceal in his diaries. For love is the dominant theme in his philosophy; hate and evil merely cast shadows on the light.

Still another aspect of the picture of Swedenborg deserves attention. Most of the reports in the literature about his supernatural gifts have practically no basis in controllable data. This applies, for example, to the greeting he is said to have transmitted to Queen Louisa Ulrika from her dead brother. There may be one exception, however: the anecdote about Swedenborg having "seen" from Gothenburg the great fire that was raging in Stockholm, which is almost five hundred kilometers away. This event occurred at the end of July 1759. Swedenborg was dining with friends; around six o'clock he grew restless, went out-of-doors and returned with the news that a fire had broken out in the south section of Stockholm, where his own house was located. When the mail coach arrived, it brought word confirming Swedenborg's vision. Even the hypercritical Kant did not question the authenticity of this anecdote, which seems to presuppose some kind of second sight and which could be verified by a number of witnesses.[165] The circumstances are certainly strange, but obviously they cannot be regarded as objective proof of Swedenborg's visionary faculty.

To those who are mainly interested in Swedenborg's works and his literary significance, the question of his mental health is scarcely of primary concern, and one is tempted to regard the psychiatric discussion as mostly a matter of terminology. From the literary point of view, it is of little help to characterize Swedenborg as a case of paranoiac hysteria or any other mental disturbance, since this in no sense explains his originality—the parallels produced from medical case histories are usually deplorably meager in comparison with the tremendous imaginative power and systematic intelligence of the seer. Furthermore, the contemptuous attitude to the alleged madman, displayed by so many of his opponents, seems both heartless and tactless, even if the diagnosis were correct, which may very well have been the case. Not least in our post-Romantic age, when so many people are desperately and pathetically looking for ways to extend the limits of consciousness, it should be more appropriate to feel empathy and admiration for Swedenborg's titanic struggle to enter into a psychical reality beyond mundane comprehension.

The Worship and Love of God

In any event, when we come to analyze and characterize the work which grew out of the fruitful turbulence of the *Journal of Dreams*, the psychiatric standpoint is as uninteresting and unproductive as in the case of the scientific works discussed hitherto. But the religious crisis was the psychological premise for the creation of the synthesis of decades of research represented by *De Cultu et Amore Dei.* Through his struggle with temptation, Swedenborg freed himself from the compulsion to carry a burdensome scientific apparatus on his shoulders, and his dream experiences gave him the courage to embark on a new literary form: his visions of the kingdom of innocence were translated into a prose poem on the creation of the universe, on the Garden of Eden, and the bliss of the first beings under God's wing.

A noble prologue introduces one of the work's basic propositions, which is the most general exposition of the *doctrina correspondentiarum:* a network of correspondences exists between particulars and the universe as a whole. The prologue thereby sets the tone for the first literary application of this correspondence theory, and it illustrates the proposition by reviving an ancient notion of correspondence, the antique belief in four ages of the world. The first chapter goes on to describe the structure of the solar system and its origin in the great egg of the universe, the *ovum mundanum:*

> There was, then, a time like no time, when the pregnant sun carried in his womb the gigantic brood of his own universe, and when, being delivered, he emitted them into the regions of air; for if they were delivered from the sun, as a parent, it is manifest that they must have burst forth from his fruitful womb. Nevertheless, it was impossible that he could carry in his burning focus, and afterwards bring forth, such heavy and inert productions, and therefore such burdens must have been the ultimate effects of his exhalation, and of the forces thence flowing and efficient. Hence it follows, that the sun was primitively overspread with effluvia excited and hatched by his real irradiation, and flowing together in abundance and from every direction to him, as to an asylum and only harbor of rest; and that from those fluids, condensed in process of time, there existed a surrounding nebulous expanse, or a mass like the white of an egg, which, with the sun included in it, would resemble the GREAT EGG OF THE UNIVERSE; also that the surface of this egg could at length derive a crust, or a kind of shell, in consequence of the rays being intercepted, and the

> apertures shut up; and this crust, the sun, when the time of parturition was at hand, by his inward heat and agitation would burst and thereby hatch a numerous offspring, equal in number to the globes visible in his universe, which still look up to him as a parent.
>
> *De Cultu et Amore Dei*, n. 9

Here the cosmological theories in the treatises of the 1710s and in *Principia* recur, this time in poetic form, but they are soon combined with the biological speculations in the *Oeconomia* series. There was eternal spring on the newborn earth, and the fertile soil bore living beings in the paradisiacal climate. This happened through trees and plants laying eggs which were counterparts of the great world egg, still another example of universal correspondences.

The second chapter, the most comprehensive in the book, deals with Adam's birth and upbringing; we shall linger at the birth scene, indeed an instance of poetic sensibility. Man too is born from an egg, carried by the tree of life in paradise and fertilized by the Lord himself. Swedenborg describes in detail, but with a sort of delicate reserve, how the soul, *anima,* builds up its body in faithful observance of the purposes for which it is intended by providence: to create a microcosm corresponding to macrocosm and thereby to link nature's exterior with her interior. When the time came for man's entrance into creation, a circle of angels guarded the precious birthplace:

> All things were now prepared; the parturient branch, according to the times of gestation, inclining itself by degrees towards the ground, at length deposited its burden commodiously on the couch spread beneath. The heavenly beings, clothed with a bright cloud, also stood by, and found that nothing had been neglected, but that all things were prepared obsequiously by nature in conformity to their provisions. Hence when the months were completed, at that time so many years, the foetus, perfectly conscious of what was decreed, himself broke through the bands and bars of his enclosure, and raised himself by his own effort into this world and its paradise, desired from the first moments of his life; and he immediately drew in with his nostrils and breast the air, which he saluted with a light kiss and which pressed in by its force as a new vital guest and spirit, for which

> the approaches and interior chambers had been previously provided, and opening by its aid a field for exertions, he excited to their offices all the powers of his body, which were already in potency and endeavor, to exercise themselves.
>
> The choicest flowers, encompassing this couch, now exhaled their odors from their deepest pores, that by them, infused into the attracted air, they might penetrate and exhilarate with rich and delicious gifts all the blood of the infant, flowing from the heart and now meeting the air. Whatever was in the kingdoms of nature, as if conscious and excited by a kind of festivity, favored, and in its own manner, greeted this birthday; for all celestial stores at this moment were effulgent, and by their influx, as it were, announced it. Choirs of the heavenly ones concluded this scene, which was the third, with the delicate vibrations of their lights, as so many tokens of gladness and favor.
>
> *De Cultu et Amore Dei,* n. 38

The sections on the growth to maturity of the first-born give concrete examples of the psychological theories in *De Anima.* In conversations with celestial tutors and his own intelligence, Adam learns of humankind's intermediate position as the recipient of divine, as well as natural light, and he hears of the struggle between God and the prince of the world and of the existential choice between love of God and love of self. But the elaborate pedagogical discourse is not easy to follow, since the author sets the entire process in Adam's own mind. A modern writer would probably have used the device of inner monologue or some other form of the stream-of-consciousness method, but Swedenborg constructed his psychological hierarchy with the help of dialogues. When Adam converses with his intelligences, Swedenborg adds a footnote to the effect that this means that he is thinking.[166]

Keeping track of the symbols requires great concentration on the part of the reader, who must also bear in mind the psychological system with its combination of rationalism and empiricism: the sensory reports are transformed by the soul *(anima)* into ideas, which are filed in their memory chambers. From these ideas, the soul then creates intelligences which, after having shared her light and the warmth of her love, are elevated to wisdoms. The mind *(mens)* is composed of these intelligences and wisdoms. The latter are in communication with the soul and thus correspond most closely to the pure intellect as described in *De Anima.* The soul, in turn, is the receiving organ for the heavenly fluid, possesses all knowledge from the beginning, and is the guiding vital principle of the body.

The second part of the work begins with a description of the birth of Eve, in which the egg notion recurs, this time combined with an allusion to the story in Genesis about her creation from one of Adam's ribs. One passage describing Eve's growth contains variations on the themes from the first part, at the same time as Swedenborg takes the opportunity to revive the psychophysical theories from the *Oeconomia* series: it is here that the dissection of the finest fibers discussed earlier are brought into the picture.[167] At the end of this part, the first beings meet one another, and the young girl is led to her bridegroom by an angel.

No further parts of this strange work were published by the author. However, there is extant the beginning of a third part, which was to deal with Adam's and Eve's life as man and wife. It opens with a description of their wedding night, marked by sublime beauty and reverence—if one reads it with the *Journal of Dreams* in mind, the old bachelor's vision of the serenity of sexual intercourse stands out as pathetic and far removed from his own experiences of *Venus vulgivaga.* Early the next morning, the young couple see a vision in the sky representing "the universe with its destinies and inmost certainties."[168]

The vision takes the form of a brilliant center of light surrounded by two girdles. The first is adorned with a host of beautiful figures which, like Adam and Eve, recline on nuptial couches and represent different kinds of love. The second girdle encloses the first one in a ring of fire, palpitated like a heart, and forms a cavity also reminiscent of the heart. Around these girdles is heaped a tremendous mass of innumerable small eggs, which are in communication with the great heart and eventually give birth to young, both human and animal. The whole is encircled by a ring of crystal in the form of an egg. After circling around its center for a long time, the crystal ring separates into strands, but these dissolve, and from them rise a human form that floats up toward the heavens. The vision ends with the circles forming spirals, which constantly create new conical figures.

Swedenborg broke off his work before he had allowed Adam to finish his interpretation of the magnificent vision, the synthesis of the synthesis and the culmination of the series of correspondences. It is quite clear, however, that the human form represents the final goal of the creation, the kingdom of spirits in the shape of an all-inclusive human body. In an earlier draft, Swedenborg had planned a work on the City of God for the year 1747, and the final vision in *De Cultu et Amore Dei* is a fragmentary realization of this project. The discussion between Adam and Eve may also give the key to the title of the book: "for APPROACH BY WORSHIP IS SUCH AS IS CONJUNCTION BY LOVE."[169] Without God's love, transmitted through Jesus Christ, we could never understand his truth and hence not worship him in a fitting manner. In the human form, the symbol of the congregation of saints, the only begotten Son is the head, the soul, and the mind.

Swedenborg sent the third part to the printers, but stopped the work after only a few pages. It is characteristic that the interruption came at the point when Adam had begun to describe the details of the vision, particularly those concerning the function of the prince of the world. It is true that the drama of Creation and the first human beings contains a number of portents of Swedenborg's development, at the same time as it is a synthesis of his scientific production; but it still represents only a short step along the road. The work is not regarded by Swedenborgians as one of the writings inspired by his illumination, and the explanation given is primarily that the author had not yet arrived at any final theological standpoint.

The belief in the Messiah as the only begotten Son of God and in a personal prince of the world is dropped in the subsequent works, as is the notion that angels and demons existed before man. In *De Cultu et Amore Dei*, Swedenborg reckoned with a fall in heaven, as described by Milton in *Paradise Lost*—the similarity is due, of course, to the common biblical source. The prince of the world and his subjects were created to function as a nexus between life and nature, but, contaminated by love of self, they revolted. The Lord then decided to destroy not only the rebels but also the entire universe. The Son, however, threw himself in the way of the Father's thunderbolts and extracted a promise that the earth would be allowed to complete its allotted span.[170] But the harmonious relationship between heaven and nature had already been destroyed, and though the prince of the world still acted as the intermediary, he did so only under compulsion. In the final theosophical system, the devil is exterminated and the role of Christ is transformed. Christ becomes an incarnation of God himself, and he descends to earth to restore man's knowledge of the Word and to correct the balance in the world of spirits, where the evil ones had multiplied and grown too powerful.

De Cultu et Amore Dei is a late example of the kind of literature usually known as the hexaemeron genre, works describing the creation of the world in six days, but it seems impossible to associate it with any specific literary model. Impulses from Milton have been cited, but it cannot be proven that Swedenborg ever read *Paradise Lost*; the similarities can be explained by the common background in the Bible and classical mythology.[171] Were one to try to specify a single significant source, the first to come to mind would be Ovid, the creator of *The Metamorphoses*, in whose writings Swedenborg had early on encountered a world of correspondences and transformational symbols. He also had witnessed how Ovid could be used in cosmogonic contexts, for example, in Thomas Burnet's *Telluris Theoris Sacra*, which gave Swedenborg the inspiration for the *ovum mundanum* symbolism. But the final impression is that Swedenborg achieved a profoundly personal synthesis of his scientific studies and his religious beliefs, an original tapestry of thoughts and impressions that had been in his mind ever since his youth, accomplished under the pressure of the forces of imagination liberated by his religious crisis.

The Years of Preparation

In March 1745, Swedenborg sent a few copies of the published parts to his friend Minister Preis in The Hague, and his covering letter indicates that he was justifiably proud of his work.[172] But the next month he experienced the vision that irrevocably diverted him from worldly learning. We know this because he told several friends about it in his old age.[173] The vision was presumably in the form of a manifestation of the Lord God himself and was a call to explain the spiritual meaning of the Word to humankind. Swedenborg obeyed the command; obviously, this did not imply any sudden or radical re-orientation. For the few years following 1745, he was absorbed in systematic Bible studies. These resulted both in extensive indices and other excerpts and in a number of attempts to base on them a presentation of their spiritual significance.

Understandably, the first effort is devoted to the book of Genesis, and this led him to evaluate his last profane work in the light of his new experiences. The outcome surprised him: the conformity was greater than he had anticipated. It should be noted, however, that Swedenborg only named the origin of the world, paradise, and the creation of Adam as points of comparison. Admittedly, Adam's birth from an egg in *De Cultu et Amore Dei* should have presented a hurdle, but Swedenborg appears to adhere to the views he expressed there; he leaves it to the reader to decide which version of Genesis he prefers, the literal or the interpretative.[174] However, this *Historia Creationis* was interrupted after the third chapter of Genesis, and Swedenborg began all over again with other interpretations.

These are aimed at determining what the Bible has to say about the kingdom of God.[175] Once more we are reminded of the working plan for 1747, *De Civitate Dei,* (On the City of God). To realize this plan, Swedenborg contemplated following typological and symbolical approaches. The purpose of the creation was to establish a society of saints to the greater glory of God, *ad majorem Dei Gloriam,* and nothing exists in this world that does not, to the best of its ability, represent this ultimate purpose. The description in Genesis of the emergence of the earth from chaos and of the progress of the days of creation thus summarizes in symbolic terms the fate of the universe and of mankind. The idea is obviously to interpret each separate verse from this basic viewpoint and then to confirm the conclusions with the pronouncements of the prophets, the apostles, and Jesus Christ himself.

With his enormous energy, Swedenborg was able to carry out a great part of his program. Many thousands of pages of his manuscript, divided into categories, are still extant, and most of them have been published by devoted disciples long after the master's death. For about two years, however, he appears to have worked only with Latin translations of the Bible; he apparently did not embark on a study of the original Hebrew text until the beginning of 1747.

He conducted the greater part of his biblical researches in Sweden after his return in the summer of 1745, concurrently with his work at the Board of Mines. There are no signs that anyone objected to the assessor's new preoccupation, assuming that it was known. Nor was it sensational in itself, bearing in mind the activities in their old age of such illustrious men as Newton. In fact, Swedenborg was recommended for promotion to a position as councillor on the Board. Instead, he chose to submit his resignation in order to have more time for his studies. In his letter of resignation, however, he merely wrote that he was engaged on "an important work." He declined any promotion, but requested that he be permitted to retain half his salary. His request was granted in the summer of 1747, and the pensioner once again departed for Holland and England.[176]

There is no need to dwell at length on these years of preparation. Swedenborg maintained the same general working habits that he followed when he wrote his scientific treatises, but the sources had been reduced. He was no longer a solitary seeker for the truth about nature and for a language that could express his discoveries with mathematical precision. Now he shared the knowledge of the spirits, was in daily communion with them, and had his views confirmed by them. This does not mean that he had reached absolute certainty on all questions. On the contrary, one finds many examples in his *Spiritual Diary* of what he regards as attempts by evil spirits to lead him astray and into temptation. But he possessed the certainty of his call, and this transformed the lost and anguished seeker into a peer of the prophets. The task remained of finding receptive ears in a skeptical world.

Visionary Scientist

In 1749, Swedenborg presented, once more in the Latin of the learned, his first reports of his voyages in the world of spirits, interpolated among exceedingly detailed expositions of the first book of the Bible. With the anonymous publication of the first volume of *Arcana Coelestia* in London in 1749, the sixty-one-year-old scholar appeared in the new role that was to render him a unique position in world literature.

7

The Secrets of Heaven

The Spiritual Tongue

In the eight great volumes of *Arcana Coelestia* (1749–1756), Swedenborg appears in the two roles that gave him fame: exegete and visionary. The evaluation of the two faces of the man varied even during his lifetime. Many were able to accept his exegesis and the rationalistic foundations of his theology, but regarded his reports from the world of spirits as embarrassing proof of his aberration. Others were repelled by his view of the Bible and of God and urged him to write only about what he had seen and heard in the world of spirits: this was the case of his first disciple in Germany, the controversial prelate F.C. Oetinger.[177]

However, it is impossible to separate the various elements of Swedenborg's literary activity after his illumination, since they form a coherent whole. Though he himself frequently felt he was writing at the dictation of the spirits, even believed his hand was guided by angels, his hermeneutics is the fruit of his years of intensive study, and his theological system also includes significant elements that had been evolved during his scientific period. The celestial topography and demography obviously have their most profound roots in the scientific theorizing on corpuscular philosophy, the monad concept, the anatomical reports of the microscopists, and cerebral physiology. With Swedenborg there is no question of stray and disconnected experiences, but rather of an overly systematized pneumatology. Evidence is thereby provided of the need for and practicability of research in his spiritual world—few literary landscapes can have been mapped with greater care and thought.

The great difference between the *Journal of Dreams* and the subsequent reports on the spiritual landscape is that, in the latter, Swedenborg relates the doings and sayings of the spirits as if they were ordinary everyday phenomena. The hesitations and doubts of the *Journal of Dreams* are almost entirely gone. There he is often uncertain how the symbolical dreams should be interpreted and finds it necessary to write down his efforts to decipher them, but in *Diarium Spirituale* we encounter the most factual and down-to-earth accounts. Paradoxically enough, the latter diary makes a more "scientific" impression, in part because the author frequently produces the most detailed physiological and psychological frames of reference, into which the language of the spirits, divine light, and celestial heat can be fitted.

A relatively insignificant expansion of the psychophysiological concepts of the *Oeconomia* series actually sufficed for Swedenborg to be able to explain in rational terms his experiences of the presence of the spirits. When in *Arcana Coelestia* he speaks of his "inner vision," which permits him to see supernatural things more clearly than everyday ones, he is obviously thinking of the inner eye of the soul (*anima*). All perception ultimately depends on *anima* as the vital principle of the body; but for information on the material world the soul depends on the organs of perception, the external ones in the eyes, ears, and skin and the internal ones in the cerebral cortex. When the body dies, *anima* is naturally freed from its boundaries and all that remains is its own vision.

This inner vision, however, is bestowed on certain favored beings, for example, the prophets, during their lifetime. Swedenborg uses the term "representative vision" for these cases, and he thereby emphasizes the connection between this and the normal intellectual process. At the same time, he draws attention to the difference between representative vision and the more unusual "living vision," which he also had known, an experience related to that described by Paul in the second epistle to the Corinthians with the words: "I know a man in Christ who fourteen years ago was caught up to the third heaven" (12:2). Lamm believed he could identify the two varieties of vision as pseudohallucinatory and psychosensorial visions, respectively, but other researchers do not agree on this point; a number of intermediate forms can probably be assumed. Swedenborg himself, however, wanted to distinguish between his daily exchanges with spirits and his very rare experiences of the other kind; it is, of course, his view that is important in this context.[178]

A great part of the spiritual diary is devoted to discussions of the language of the spirits. Normally, Swedenborg clearly comprehended their speech with his "inner hearing" whose function corresponds to that of the inner vision; he analyzed it with the help of the tremulation model he had developed back in the 1710s. He frequently affirmed, most categorically in *De Anima*, that thought is a kind of inner speech, and this theory is justified by what happens in the material substratum of thought. From that point of view, thought comprises changes of conditions in the brain, which produce undulatory movements in the spiritual fluid; and these, in turn, bring about movements in the organs of speech, which result in corresponding tremulations in the air, that is, sounds. When Swedenborg became aware of the spirits' speech, a similar process was set in motion but without any external influence. The speech of an angel or a spirit influences thought and then acts on the hearing organs from within, according to *De Coelo* (Heaven and Hell, 1758).[179] Consequently, the sound of this speech is as distinct as that of a human being, even though it can only be heard by one who has been exposed to this inner influence.

This physiological model can also be combined with the doctrine of forms in order to define how the influence can be brought about, and here we have an exceptionally good illustration of Swedenborg's talent for coordinating his pneumatic experiences with the metaphysical system he had already developed, which, conversely, had served as a preparation for such experiences. In his discussion of the serpent's conversations with Eve in *Historia Creationis*, he also goes into the speech of the inhabitants of heaven and states that this is produced through marvelous circumvolutions of the celestial form.[180]

This agrees well with what he wrote about this supreme form of nature in his work on fibers. It also explains the relations between the speech of the spirits and a person's normal tongue. The speech of the spirits is universal; hence, it corresponds with every individual mortal tongue, since all words have an inner significance corresponding to the celestial form. A basic concept that recurs constantly in *Diarium Spirituale* is that each one of our everyday words can have thousands of inner counterparts. Swedenborg justifies this idea in characteristic fashion by pointing out how tremendously the microscopists expanded our knowledge when their vision was intensified by ground lenses; an equally unbelievable expansion will be the result of the theory of correspondence.[181]

By comparing words and thought with the doctrine of forms applied psychophysiologically, Swedenborg arrived at the conclusion that language in the world of spirits is an inner speech and that the spirits thus communicate through direct transference of thoughts. This is also the way in which he imagined the conversations between the first human beings and the spirits before the relationship was broken by the fall of man. When spirits now converse with living beings, the recipients comprehend what is said in their own language; the explanation is that the words have a universal inner significance, which is all that the spirits can understand, while human beings can and must transpose it to their own particular level. Certain traces of these universal meanings can be found in our normal speech, for example, when we use various kinds of images—*videre,* see, for understand, grasp, etc.[182]

In *Diarium Spirituale*, Swedenborg is greatly concerned with noting and analyzing his experiences of the speech of the spirits, but since it was written parallel with intensive studies of the Old Testament, it is not surprising that the problem of written language is also discussed. The more familiar he became with Hebrew, the more he tended to put the language of the Word in a special class. In his view, the reason that Hebrew letters are best suited to convey the secrets of heaven is that there is a correspondence between their form and that of the heavenly influx. In ancient times, according to Swedenborg, this correspondence was total, but in later Hebrew writing, it applied only to the softly rounded characters.[183]

Conceivably Swedenborg was inspired to some extent by Caspar Neumann, the theologian referred to by Wolff, who presented a theory on the hieroglyphic nature of written Hebrew at the beginning of the eighteenth century. Ever since ancient times, Hebrew philologists have called the elements of letters *jodim* and have regarded them as combined by different variations of the character *jodh*. Neumann considered that *jodh* is the sign of extension and, on the basis of his Cartesian conviction, he explained that extension is an attribute of the material bodies. Consequently, the Hebrew characters represent *imagines rerum corporearum*, images of material things. Neumann also stressed that the Hebrew system of writing vowels separately corresponds with the absolute separation of soul from body, "discovered" by Descartes.[184]

There is no question that Swedenborg's view of the correspondence between Hebrew writing and the celestial form produced the same effect as Neumann's, and his view of the special position of the vowels is also the same. In *Arcana Coelestia*, Swedenborg carried out a preliminary classification of the meanings of words according to the vowel frequency in Hebrew. The first three vowels are usually found in words belonging to the spiritual class, while the last two are generally present in words of the heavenly class. In this way, he arrived at a hermeneutic principle that worked mechanically, and this must have been particularly important when he was faced with translating the enormous volume of personal and place names in the books of Moses. The same classification recurs in somewhat different form in *De Coelo*, when he describes the speech of the angels and spirits. In the case of spiritual angels, words are related to *E* and *I*, while the speech of heavenly angels resound in more powerful harmonies with the help of *O* and *U*. Their intervention in the speech of humans is also determined by this distinction, which thus allowed Swedenborg quickly to identify his interlocutors.[185]

Swedenborg's stories of the spirits' speech are obviously obscure and often contradictory, particularly with regard to their intercommunication. It is clear, however, that they were determined by the metaphysical frame of reference from his scientific stage. In the same way as innumerable muscles and fibers respond to a simple action, host of angels and spirits correspond to every thought, according to an early entry in *Diarium Spirituale.*[186] The doctrine of forms expressed in the *Oeconomia* series is applicable in both cases, and, in the most profound sense, the communication of the spirits is a counterpart of the universal language, which he tried so desperately to grasp during his last scientific years. With the entrance into the circle of spirits, he no longer felt the need to construct such a language artificially. For it is there, concealed within each word; and an interpretation of the Holy Scriptures in accordance with the *doctrina correspondentiarum* provides humanity with an adequate understanding of the language that was our original tongue.

The links with the past are thus very clear, and they are not concealed by the concrete linguistic problems that force themselves on the exegete. Hebrew is given a place of honor because its writing has a hieroglyphic import, at least to some degree, but it is never regarded as a particularly holy language or as the heavenly tongue. Swedenborg was not attracted by the Kabbalah, and we find none of Boehme's linguistic mysticism in his writings. He was and continued to be marked by his "immoderate desire" for mathematics, and the beauty possessed by his spiritual world is universal and hierarchic but not individual. Herder once complained that all the great historical personalities whom Swedenborg encountered in the spiritual world spoke in the same way, i.e., like Swedenborg himself.[187] This is a legitimate criticism, particularly on the part of a man who considered the individual personality to be of supreme importance. Swedenborg, however, did not belong among the romantics but among the rationalists, among the worshippers of geometry and mechanics.

Naturally, this does not mean that many of Swedenborg's memorabilia are not of high literary standard; nor that more recent romantics than Herder were fascinated by his thoroughly symbolic view of the universe that resulted from the theory of correspondence. A system with such pretensions to totality as Swedenborg's must inevitably contain incompatible elements, not to say direct contradictions, and this is particularly applicable to supernatural conceptions. He affirms time and again what he has seen and heard, but a comparison of the unedited notes in *Diarium Spirituale* with the well-disposed and well-founded memorabilia in the published works reveals a striking difference; there can be no doubt that he worked hard to achieve a literary effect that is completely lacking in the raw material.

These efforts naturally provide an explanation for some obscure passages. With respect to the speech of the angels, it is frequently stated that their intercommunication represented direct thought transference without any outside medium, something that also applied to the communications between the first human beings and the spirits. In the memorabilia, Swedenborg frequently has the angels speak with one another in a human tongue, and he even is permitted to read letters written in Hebrew and Latin characters in the world of spirits.[188] The diary contains examples of "spiritual language words," which do not exist in any human language and in which each sound represents a whole concept; words of this kind were never mentioned in the published writings.[189]

The reader also frequently loses his or her bearings in the celestial topography, particularly if one is not sure at which level of ambition one finds oneself. It is impossible with the space available to analyze in detail the evolution from the somewhat hesitant, unadorned narrative style of *Arcana Coelestia* to the final artistic confidence of *De Amore Conjugiali* (Conjugial Love, 1768), which was perhaps Swedenborg's greatest work; but we can study a few typical examples.

The Structure of the Spiritual World

The first published stories of the world of spirits explain logically and instructively enough how the soul is released from the body at the very moment of death. They begin in the third chapter of *Arcana Coelestia*, and are based on Swedenborg's observations in a kind of experimental situation, obviously close to the "near death-experiences" which have garnered much attention in recent years.[190] Two celestial angels occupy the heart, and two others sit by the head. A person's thoughts are preserved and guided by the angels at the moment of death, and they stay with the soul until it has reached the shore of eternal life, normally in the form of the world of spirits, in which the choice of final abode is made. When the celestial angels depart after completing their task, they are replaced by spiritual angels, who open the inner eye of the soul. Until that point, the soul has only been able to experience its liberation intellectually, but now it can perceive it sensuously.

At this point, the liberated soul may itself seek associations with the spirits. For it is a basic element in Swedenborg's conception of eternal life that every spirit gravitates toward the circle of spirits that share its thoughts and feelings as they were formed during its mortal life: as already suggested, this is one reason that Swedenborg attributed such importance to the right indoctrination. Naturally, this is a question of innermost thoughts and inclinations. A person's masks are dropped when he or she is faced with the divine light, and the exposure occurs in the most effective manner: the spirit in question is unable to remain with others than its peers; all others trouble it by their very presence.

When the spirit has found its right abode, it lives for a time a life that is in full agreement with its life on earth. This is possible through the doctrine of correspondence, which allows the spiritual territory to be an exact counterpart of the terrestrial. The agreement is so complete that many "newly dead" do not discover for a long time that they have departed their mortal life, but eventually they begin their new life from this intermediary stage. Swedenborg speaks unreservedly of "time" in these narratives; but as appears from his concurrent biblical interpretation, he does not regard it as a time concept but as a condition. The same applies to space and its changes, but he must, like the Bible itself, speak to humans after the manner of humans.

For the truly evil spirits, the road from the interim world leads to hell, while those who lived in any degree of faith in the Lord are led to heaven by degrees. It is scarcely necessary to add that Swedenborg provides no exclusively Christian definition of these faithful ones. The process may be very protracted and require considerable guidance, but Swedenborg also observed cases in which spirits were taken immediately to heaven. Their worldly misconceptions of heaven and celestial joy are thoroughly revised. It is not a matter of ruling others from a high place. Heaven is not a solemn ceremonial hall into which a few are admitted; nor does it consist of a life of ease or of constant praising and celebrating the Lord. Neither is one person's heaven or hell identical with another's in joy or in torment. Swedenborg learned from Leibniz's infinitely varied monad harmony that each soul has its own other-world existence. The spirits form societies, which are in perfect harmony within themselves and with others, and the good societies are ultimately joined in a universal unit, which has human form, *Maximus Homo*.[191] Thus, each spirit becomes an infinitesimal corpuscle in this vast body, a counterpart of the tissue particles that Swedenborg had heard of with humble reverence from the microscopists, and the universal figure is permeated by divine life just as the spiritual fluid is supreme in the human body.

Vigor and activity characterize Swedenborg's spiritual world, regardless of the perspective from which he happens to treat it. In the very first memorabilia in *Arcana Coelestia*, he emphasizes that the life of the angels consists of working for useful ends and of charitable deeds. This means, among other things, that they combat the influence of the evil spirits on living beings and awake in humans the desire for goodness and purity. Another angelic task is also to instruct newly arrived spirits about heaven and its wonders; it is here that angels find their greatest happiness, in which respect they are in agreement with their chronicler. Some of these functions, distorted in the cracked glass of evil, are also performed by the inhabitants of hell.

Every living being, though unaware of it, belongs to some society of spirits in the sense that he is united with at least two good and two evil spirits which balance each other's influence.[192] According to Swedenborg, man could not exist for a moment without these transmitters of the spiritual vital fluid; and he also associates this spiritual nexus with the frequent biblical references to the Lord speaking to humanity. But the spirits of hell are not a part of the Grand Man,

Maximus Homo; they are in truth rejected, and their evil is not permitted to reach beyond carefully drawn boundaries. If these are exceeded, the diabolical crew is called to account; later on we shall see how Swedenborg envisaged this process.

The section introducing chapter 11 of Genesis describes Swedenborg's conception of time and space in the spiritual world and of the positions occupied by *Maximus Homo* in relation to the one truly fixed point of the universe, the Lord God.[193] The spiritual societies are imagined to be separate in space, even though space does not exist in reality. Their positions and distances are established in relationship to the observer, which means that if one society is to the right, it will always be seen on the right, regardless of changes in the position of the observer's body. If one thinks of a spirit, it will immediately appear no matter where it has been previously. When Swedenborg speaks of space and time in his stories of the spiritual world, which he does constantly, particularly in his later works, he thus takes full advantage of poetic license. His defense for this is not only pedagogical, but is also based on illustrious examples: the lower spirits are said to experience space and time in the worldly sense.

With regard to the positions of angels and spirits, it is stated in *Arcana* that the angels are on the right side of the Lord, the evil spirits on the left—the pattern is, of course, set in the Word—the "intermediate" variety in front, and malicious spirits behind the Lord; the high-minded are above his head, and the torture chambers of hell are under his feet. This order is immutable and, as we have seen, independent of the position of the observer. The grouping forms a *Maximus Homo*, and the reason for it is the omnipresence of the Lord. The universal spiritual being is the visualization of the Lord's thought, a "materialization" of *l'ordre immuable*, the unalterable order, which, as Swedenborg had learned from Malebranche, also applied to the Creator himself.

Interpreted in this manner, Swedenborg's philosophy in general and his spiritual vision in particular make an extremely static impression. This may well be justified when one considers how he may have understood the original divine conception and its fulfillment, his attempts to concretize God's eternal and actual presence in categories strongly influenced by Leibniz and Malebranche. But he also had to include the cosmic time dimension in his system, even if it did not apply in the spiritual world. The Grand Man naturally must be regarded dynamically, as something that is constantly being perfected with the arrival of new souls and with the transpositions of spirits; for Swedenborg did not believe in pre-existence in any sense other than that of God's original conception, his providence, and his prescience.

Indeed, spirits who awaken in the intermediate stage are in constant search for their right societies in the body of the Grand Man. It is true that this search should not be regarded as a pilgrimage from one place to another, but rather as a change of condition: at the same time it represents a dynamic process. In his efforts to make eternity comprehensible to human beings, Swedenborg actually pursues his way of thinking from *De Infinito,* and even then he was fully aware that no one living in the flesh can do more than confound the infinity of God with that of space and eternity with time *ad infinitum.* But he no longer fears these misunderstandings as he had in 1734, and instead makes the greatest virtue of necessity. He clearly relies on his general reservations, which are frequently repeated and must be regarded as an instruction to the reader to transpose his memorabilia with the help of the doctrine of correspondence.

Maximus Homo is the all-inclusive form; but the most diverse scenes are included within its framework, and the seer sometimes specifies their position in the body of the universal being. Swedenborg relates in *Arcana* that he once was led to a paradisiacal garden, which was situated "slightly above the corner of the right eye," and there he found the Eden of eternal spring as he had envisaged it in the works of his youth, with a varied and magnificent wealth of symbolical things:

> Certain souls lately deceased, who, in consequence of the principles they had imbibed in the world, doubted the possibility of such things existing in another life, where there is neither wood nor stone, being taken up into that paradise, and discoursing thence with me, said in their astonishment, that what they saw was inexpressible, and that they could not represent its inexpressibility by any idea, and that delights and happiness shone forth from every object, and this with

successive varieties. The souls that are introduced into heaven are generally first conducted to such paradisiacal scenes. But the angels behold such things with other eyes, not being delighted with the paradises, but with the representatives, and thus with the celestial and spiritual things which give them birth. It was from these celestial and spiritual things that the most ancient church derived their paradisiacal scenery.

Arcana Coelestia, n. 1622 [2]

Spirits from Other Planets

In *Arcana Coelestia,* Swedenborg also gives a systematic presentation of the spiritual societies in *Maximus Homo*, and he does not deny himself the pleasure of reaching out to the other planets of the solar system and even farther into space.[194] So tremendous is the universal being that its body cannot be constructed by terrestrial inhabitants only, but requires a population spread through cosmos. We thus encounter in the spiritual world Swedenborg's version of the ancient dream of space travel, Lucian's, Ariosto's, and many others' planetary visits, which were revived in Fontenelle's *Entretiens sur la pluralité des mondes*—published 1686 but translated into Swedish in 1759.

Swedenborg found it unreasonable that the tremendous planetary masses should be uninhabited and exist merely as sources of light for the inhabitants of the earth. This would be in conflict with the divine principle of use, which governs cosmos. He therefore goes through the planets, one after the other, and tells about meetings with spirits from them, after which he moves on to the firmament of fixed stars. The best of the planetary spirits originate in Mars. Their speech requires no external sound, but reaches the brain through an internal passage, via the Eustachian tube, in the same way as he believed was the case of the people of the most ancient church on earth. In *Maximus Homo*, whose organic correspondences were discussed at length in earlier sections of *Arcana Coelestia,* these former Martians constitute the intermediate province that connects the cerebrum with the cerebellum.[195]

The inhabitants of Jupiter receive the greatest attention. Their principal characteristic is integrity, and their social system and form of life, with its idealized communistic large families, correspond to those of the most ancient church. This means that they also correspond to the ideal state of the Golden Age. Their speech is exclusively physiognomical: that is to say, the face mirrors the soul without pretense. This, too, is in agreement with the original state on earth, and the origin of all languages is thus mimetic; words were invented successively but not immediately conveyed to Adam, as many scholars have believed in the history of linguistic theory. In the Grand Man, these spirits from Jupiter represent the "imaginative power of the mind."

Swedenborg's scientific lack of prejudice allowed him to present undaunted the most absurd information about the lunar spirits. The inhabitants of the moon speak not with the help of their lungs but from the abdomen in eructations—this odious phenomenon is called by its medical name—and the reason is difficult to contest: the moon, unlike other "earths," lacks its own atmosphere, and the lungs can therefore not function as a source of the tremulation of the vocal cords.

He also makes a number of calculations concerning the space needed for three hundred million beings from each of one million planets and stars for two hundred generations. No one can complain about lack of concreteness on the part of the astronaut Swedenborg, nor accuse him of tellurian limitations. Still, Earth remains the center of the universe in the most essential sense, since the Lord let himself be born here. The main reason for this, according to Swedenborg, is that the art of writing is older here than on other planets, if it even exists elsewhere.[196] This gave the inhabitants of Earth better instruments to retain and disseminate the Word than other beings. Few arguments illustrate more clearly the extremely literary and intellectual nature of the inspiration that animated Swedenborg.

The Last Judgment

The many scattered reports on space at the end of *Arcana Coelestia* were assembled and put out in a small volume in 1758.[197] At the same time, Swedenborg published four minor works in London. One of these deals with the Last Judgment and deserves special attention as an example of the evolution of Swedenborg's theology during his work on biblical texts. One day in 1748, he noted in *Diarium Spirituale* that he had seen the figure 57 in a vision.[198] This was neither the first nor the last time that he received this kind of numerical revelation, but on this occasion he combined it with his thoughts on the true meaning of the Last Judgment.

Early in *Arcana,* he rejects the usual interpretation that the Day of Judgment represents the end of the world.[199] To him it was quite clear that the doomsday trumpet would proclaim the end of all churches, the stage at which faith had disappeared. The Deluge was the Last Judgment of the oldest church, and the doomsday of the Israelitic church corresponded with the arrival of the Lord on earth. Certain adjustments in Swedenborg's writing of history occur later, but the basic pattern remains unchanged. The Christian church, which succeeded the Jewish, underwent a progressive deterioration; in his last work, he emphasized the Council of Nicaea (325 CE) as a fateful turning point through its formulation of the Trinity dogma. A judgment now awaited degenerate Christianity, and Swedenborg himself witnessed the destruction of what he calls Babylon in the year 1757—here we have the explanation of the vision of the figure 57.

What then happened was confined to the world of spirits, and the story is told in *De Ultimo Judicio* (On the Last Judgment), a small treatise published in the productive year 1758. The spiritual societies doomed to extinction were peopled by spirits that had lived outwardly impeccable lives and had therefore been permitted to group themselves in the outskirts of heaven. However, due to the miserable state of faith of the churches, these spirits grew so numerous that they threatened to destroy the connection between heaven and earth.

Swedenborg specifies several confessions among these spirits and also different nations. Those in the worst predicament are the Roman Catholics, and it is their confession that he identifies with the Babylon of the Apocalypse. More precisely, Babylon refers to all who want to rule with the help of religion. And this, according to Swedenborg, is the aim of the Roman Catholic Church, whose most unforgivable crime is that it conceals the Holy Writ from the layman. Bearing in mind his extreme biblicism, this is an understandable attitude, and the sentence he hands down is in proportion to the crime. Earthquakes, tempests, and conflagrations annihilate proud Babylon, and the inhabitants are hurled into a lake of black water. The entire area is engulfed in a black cloud, which takes on the shape of a dragon, the symbol of the falsity of the religion. It is obvious that the colors and shapes of the lugubrious landscape originate in the Revelation of St. John, to which Swedenborg was to devote his hermeneutical labors during the next decade.

In *Vera Christiana Religio* (*The True Christian Religion*, 1771), Swedenborg still retains his description of the Last Judgment in the world of spirits of 1757, but his interest is now centered primarily on an explication of the words promising the Second Coming of the Lord.[200] This will not happen *in persona* but through the Word, through humanity's regaining an understanding of its spiritual significance. The one who transmits this understanding is Swedenborg himself, and with his emergence was founded the New Church, the church known as New Jerusalem in the Apocalypse. Swedenborg's ponderings on the Last Judgment finally resulted in a definitive date. On June 19, 1770, after the completion of *Vera Christiana Religio*, the Lord called together the twelve apostles, who had once accompanied him on earth, and sent them into the world of spirits to preach the gospel according to Emanuel Swedenborg.[201] Naturally it is not expressed in that way, but the intent can scarcely be interpreted otherwise. His fidelity to the call of the year 1745 received its highest conceivable reward in the last memorabilia published during his lifetime.

The Seer as an Artist

Hitherto, we have dwelt on the content of Swedenborg's stories of the world of spirits. However, they also have a literary side, which is of considerable interest, particularly in the later works. To judge these qualities fairly, one must naturally proceed from the unlimited and yet disciplined opportunities for symbolical descriptions, which the doctrine of correspondence offered its author. As far back as *De Cultu et*

Amore Dei, Swedenborg had created paradisiacal scenes in which animals and plants represented symbols, flowers "which were never afterwards seen, namely, which had inscribed on their leaves, and presented to view in different ways, the series of the fates of the globe and the nature of the universe" and birds which "bore the marks of paradise itself, or of its grand scenery, in their feathers" (n. 19, 26). The interior of Adam's intellect is described as a hearth of gold and diamonds on which a fire burned clear and strong (n. 51, 67): the significance is explained in a later passage in a manner closely reminiscent of the earliest exposition in *Clavis Hieroglyphica*.

In the work on heaven and hell published in 1758, *De Coelo*, which summarizes what was said in *Arcana Coelestia* in connection with the interpretation of the first two books of Moses, the doctrine of correspondence is exemplified by the animals mentioned in the biblical text:

> The animals of the earth, in general, correspond to affections; tame and useful animals to good affections; fierce and useless animals to evil affections. In particular, oxen and bullocks correspond to the affections of the natural mind, and sheep and lambs to the affections of the spiritual mind; but winged creatures, according to their species, correspond to the intellectual things of both minds; and hence it is that various animals, as oxen, bullocks, rams, sheep, she-goats, he-goats, he-lambs, she-lambs, pigeons, and turtle-doves, were accepted for holy use in the Israelitic church; for that church was a representative church, and those animals were used as sacrifices and burnt-offerings. For in that use they corresponded to things spiritual, which were understood in heaven according to the correspondences.
>
> *De Coelo* (Heaven and Hell), n. 110

But the heavenly symbols do not only consist of flowers and animals but also of artificial objects: houses, palaces, cities. Where the angels live in societies,

> their habitations are contiguous one to another, and arranged in the form of a city, with streets, ways, and squares, exactly like the cities on our earth. I have been allowed to walk through them, and to look

> about on every side, and occasionally to enter the houses. This occurred when I was in a state of full wakefulness, and my interior sight was opened.
>
> *De Coelo,* n. 184

And the palaces, with their building materials copied from the New Jerusalem of the Apocalypse and their colors the utmost perfection of art, comprise the archetypes of the centers of power which Swedenborg had admired on his travels to Copenhagen, Hanover, Paris, London, and the Italian cities from Milan to Rome.

The opposite to these heavenly dwellings are infernal localities, each one of which corresponds in its own way to the habits of its inhabitants, and their dominating love—*amor regnans* is Swedenborg's basic concept when he tries to explain the distribution of the spirits between the innumerable levels in the divine order. Those who had been intriguing and treacherous live in dark caves, those who had practiced the sciences from intellectual pride live in sandy and sterile places, adulterers spend their days in hideous brothels (*De Coelo,* n. 488). Swedenborg also witnessed the dwellings of the most evil: their holes and caves in barren mountains, their lairs and dens like those of wild beasts in somber forests, their miserable cities of charred houses and hovels huddled together in alleys:

> Within the houses infernal spirits are engaged in continual quarrels, enmities, blows, and fightings; in the streets and lanes, robberies and depredations are committed. In some of the hells there are mere brothels, disgusting to the sight and filled with all kinds of filth and excrement.
>
> *De Coelo,* n. 586

The reports from heaven and hell thus describe the "representations and appearances" that form the environment of the spirits, and are suggestive enough even in *Arcana Coelestia* and the immediately succeeding works. But the most detailed and best written memorabilia were presented by Swedenborg in the remarkable treatise *De Amore Conjugiali* (Conjugial Love), published in Amsterdam in 1768 with the author's name on the title page, unlike all the previous works by the seer. In the ten years between this work and *De Coelo*, Swedenborg had penetrated the Revelation in the same extremely detailed manner as he had interpreted Genesis and Exodus in *Arcana*, and this exegetical study had a great influence on the new descriptions of the world of spirits. They are much more individualized and concrete than before, and one feels that the author presents them with greater confidence and a stronger aesthetic aspiration.

De Amore Conjugiali seems to occupy the same place among the theosophical works as *De Cultu et Amore Dei* among the scientific, a sort of synthesis in a freer spirit of inspiration. That Swedenborg introduces the book with a detailed description of the world to come seems almost to be a declaration of intent and a positive response to those who, like Oetinger, wanted him to write only *ex auditis et visis*, from things heard and seen. In any case, these memorabilia represent the apex of his theosophical writings, and most of them also recur in the work that Swedenborg intended to be his spiritual testament, *Vera Christiana Religio* (1771).[202]

De Amore Conjugiali opens with an image from the general storeroom of Christian art. An angel is in flight with a trumpet at his lips, garbed in a mantle flowing in the wind—we recognize the harbinger from an abundance of church paintings and ultimately from the Apocalypse. The angel descends to Swedenborg, who is in the world of spirits, and informs him of his task. The trumpet calls convoke those Christian spirits known for their wisdom and intelligence for a debate on the subject of celestial joy and eternal salvation; the reason is that newly arrived spirits had told of the total ignorance of these questions on earth. The angels in one society now want to unearth the truth for themselves, and the harbinger invites Swedenborg to listen to the conference. After half an hour, two groups arrive from the north, the south, and the west and one from the east. The last group was invisible to the others, however, due to its powerful radiation—as we recall, the Lord himself is always in the east. The groups collect in a circle in the conference hall erected for the occasion and begin to present their views.[203]

This amusing situation is just as reminiscent of our own conference-ridden times as the views of the groups are familiar, but the great difference lies in the narrator's richness of detail and literary setting. For each of the deluded groups is permitted to experience a "materialization" of its notion of heaven. The spirits from the north, who believed that celestial joy consists of an unending witty conversation in the full sense of the word, are ushered into a great house of more than fifty rooms, in which various kinds of conversation are going on.

Some of the groups are exchanging the news of the day, others are talking about women, court gossip, politics, business, and literature, about churches and sects—in other words, a conglomeration of what a man of the world like Swedenborg apparently would consider representative of the fashionable drawing-room culture of the day, a counterpart in the world of spirits to the European bourgeois society of the Habermas eighteenth-century model. When he is allowed to enter the house, Swedenborg encounters a familiar scene: some of those present are eager to talk, others long to ask questions, a third group wants to listen and learn. But many of them are in a state of desperation and crave escape; they are exhausted after three days of social life. All exits are closed, however, and their entreaties are answered with the cruelly ironical words: "Remain and enjoy the joys of heaven" (n. 5).

The merciful angel releases the captives from the curse of conversation, after having, in the manner of the teacher, pointed out to them the lessons of their experience. Celestial joy is incompatible with a life devoted to pleasure; instead, its absolute prerequisite is usefulness, the fulfillment of social tasks.

The same theme is varied in the next group, which believes that celestial bliss consists of endless eating and drinking in the company of the patriarchs and the apostles. The feast described by Swedenborg competes in opulence with Trimalchio's in Petronius. The guests are forced to recline at the tables with one after another of their fifteen hosts, after which they have to begin all over again.

However, the whole scene is only a pretense, in which the parts of the patriarchs are played by equally deluded spirits, though of lower social origin and probably of sectarian upbringing:

> The leaders whom you saw at the heads of the tables were counterfeited Fathers, men largely from the peasantry, who being bearded and comparatively wealthy are rather puffed up, on whom, too, the phantasy has been induced that they are those ancient Fathers.
>
> *De Amore Conjugiali*, n. 6

The guests are soon nauseated by all the food and try to get out, but the guards stop them, too: "Had they eaten yet with Peter, and with Paul? They were told that they should be ashamed to leave before they had done so; it would not be proper" (n. 6).

Those who had imagined heaven as a paradisiacal garden of pleasure are also seized with disgust; but the worst predicament befalls those for whom bliss was an uninterrupted divine service. Swedenborg's description of their appearance after two days of constant worship bears the stamp of personal experience, not so much through the reasons he gives but because he had spent a long life in the sermon-prone Lutheran tradition—at this stage, he must have spent a total of at least half a year listening to sermons and homilies. One is almost tempted to interpret his wildly comic description as a revenge for past sufferings. The angel shows the audience:

> They looked, and most were asleep, and those who were awake yawned again and again. Some—what with the constant uplifting of their thoughts to God, without consideration for the body—looked like faces severed from the body. They felt so themselves, and hence appeared so to others. Some were wild-eyed from gazing into space. In short, their breathing was labored, and they were wearied in spirit from the tedium. They had their backs to the pulpit, and were shouting, "Our ears are stunned. Bring your sermons to an end. We no longer hear a word, and begin to loathe the sound." Then they arose, rushed in a body to the doors, broke them open, jostled the guards and drove them off.
>
> Seeing this the priests followed, and clinging close to them, went on teaching and praying, sighing and exhorting, "Keep up the service! Glorify God! Sanctify yourselves! In this forecourt of heaven we will initiate you into everlasting glorification of God in a magnificent and most spacious temple in heaven, and so into the enjoyment of

> eternal happiness." But these appeals went uncomprehended and were hardly heard, so dull were their minds after the two days' inactivity and detention from domestic and business affairs. Still, when they tried to get away, the priests seized them by the arms and sleeves, to push them into the buildings where they were to preach; but in vain. The people cried, "Let us alone. We shall swoon."
>
> *De Amore Conjugiali*, n. 9

After having staged these scenes of concrete pedagogics in the spiritual world, the angels open the door of heaven to a group of ten favored ones, and they—including Swedenborg—experience the celestial realities, which had been grossly caricatured by man's dull imagination. Here they are led to a palace of incomprehensible beauty:

> It was large, built of porphyry with a foundation of jasper, and had six lofty columns of lapis lazuli at the entrance. The roof was sheets of gold; the lofty windows were of clearest crystal with frames of gold. They were then conducted inside and led from room to room, where they saw ornamentation of ineffable beauty, and, decorating the ceilings, inimitable carved work. Along the walls stood silver tables inlaid with gold on which were various useful articles of precious stones and of whole gems in heavenly patterns.
>
> *De Amore Conjugiali*, n.12

The palace is set in a park in which the trees are arranged in a way which seems to allow for endless extension; obviously, he refers here to what he called in *Oeconomia* the perpetuo-spiral form, the form of vortical movement that is the highest that can exist materially. The visitors are invited to meet the prince of the angelic society and to sup at his table; and Swedenborg takes the time to describe in detail the garb of the prince and the members of his court, a peculiar example of his ability to combine acute memories of observed facts with symbolical fantasies:

> He was dressed in a long purple robe, embroidered with stars of the color of silver. Under the robe he wore a tunic of shining silk of a blue shade. This was open at the breast, where the badge of his

> society was to be seen on a sash. The badge showed an eagle on a treetop, brooding over her young; it was of shining gold in a circle of diamonds. The chief counsellors wore garments not unlike those of the prince but without the badge, in place of which carved sapphires hung from the neck by a gold chain. The courtiers wore gowns of brown, on which were woven flowers around young eagles. Their tunics were of an opal-colored silk, as were their breeches and stockings.
>
> *De Amore Conjugiali*, n. 15

The guests are also allowed to converse with wise men about salvation; for three days, they live the heavenly life in the city by the palace: they are invited to a wedding, representing the marriage between the Lord and the church, which also signifies the union of love and wisdom; and finally they take part in the heavenly celebration of the sabbath and hear an inspiring sermon on the Word.

We cannot dwell longer on this introductory memorabile. It ends with a series of references to the books of Revelation, Ezekiel, and Daniel, which obviously were the principal sources of inspiration for the celestial scenes. To this biblical background should be added Swedenborg's classical learning, his obvious fascination with emblematics as evidenced by the eagles, and his familiarity with life at court and in palatial surroundings.

Marriages in Heaven

De Amore Conjugiali treats marital love, and the first chapter introduces the problems. It is symptomatic that the first memorabile describes a wedding in heaven and explains its symbolical meaning: it gives the key to the continuation. The doctrine of marriages in heaven plays a central role in Swedenborg's theosophy; but it remains a controversial and, to him, a problematic role, especially in view of Christ's reply to the Sadducees in all the synoptic gospels that there will be no marriages after the resurrection (Matthew 22:23–32; Mark. 12:18–27; Luke 20:27–38). Swedenborg solves this problem in his usual way by referring to the spiritual meaning of the biblical texts, which in this case would be that Christ meant humankind's spiritual union with the Lord, which occurs during life on earth.[204]

Celestial marriages are obviously a logical sequence of his anthropomorphic concepts of the beyond, which, in turn, go back to the doctrine of correspondence. But his gospel undoubtedly has more profound personal roots. The *Journal of Dreams* bears traces of a strong sexuality, and he frequently expressed great interest in the physiological and psychological problems of sex in his scientific writings; his experiences in *le beau monde* must also have brought the moral question into harsh and painful focus for a man with his reverence for the divine order. It is scarcely surprising, but perhaps pathetic, that the elderly bachelor gave marriage, in the full and rich sense of the term, such a central position in his vision of the land of beatitude. And a sentimental tradition has it that he mentioned by name a lady of high degree as his future wife in heaven; even if untrue, it is probable that his exquisite descriptions of married couples in heaven were influenced by his own dream of happiness.[205]

At the same time, it would be quite wrong to portray these heavenly scenes as sentimental. On the contrary, they are characterized by the same delight in factual data and the same peculiar objectivity as always, and in addition often by a tolerant humor, which is sometimes irresistible. For example, he tells of three newly arrived spirits, two of whom were young men bursting with sexual power and desire. They are told that, although there is love between the sexes in heaven, it is chaste and has no resemblance to mundane lust. The young men are crushed:

> "Just what is sexual love without allurements?" the newcomers ask. As they thought about this love, they sighed and exclaimed: "Oh, how insipid is the joy of heaven! What young man can desire heaven then? Is such love not empty and lifeless?"
>
> *De Amore Conjugiali*, n. 44 [3]

Their spiritual mentors laugh at their reaction and the naked sexual appetite it reveals. They explain how the pure and chaste intercourse between the sexes is filled with inner joy and causes the mind to expand. They go on to say that the chastity of newcomers to heaven is tested by leaving them alone with young girls,

> who perceive from the tone of voice and from the speech, face, eyes, bearing, and outflowing sphere, what their quality is in respect to love for the sex; and if it is unchaste they flee away, and tell their companions that they have seen satyrs or priapi.
>
> *De Amore Conjugiali,* n. 44 [4]

This, of course, further chills the ardor of the curious playboys, who continue their complaints. The spirits now become irritated by their childishness and reprimand them. In heaven, the only sexual love is between spouses, and what the young men have in mind is clearly a prodigal sowing of wild oats. This, however, is literally unthinkable in heaven; but all the delights of love can be enjoyed in marriage. No children are conceived, only spiritual offspring of love and wisdom, which results in happy consequences, interpreted by Swedenborg with the help of his old physiological model:

> Hence it is, too, that angels do not become sad after the delights, as some do on earth, but cheerful. This results from the perpetual influx of fresh powers succeeding the former, renewing and enlightening the angels. For all who come into heaven return into the springtime of their youth and into the vigor of those years, and remain so to eternity.
>
> *De Amore Conjugiali,* n. 44 [9]

This last feature was stressed in *Arcana Coelestia* as a feature of heavenly bliss: there would be no ailing old men or faded old women, but only young people in the prime of life.[206] When the dead reach the interim stage, they may sometimes meet their husband or wife and live together for a time. The more detached they grow from their mortal selves, the clearer becomes the true nature of their union. If they discover that they have not loved one another with true conjugal love, they are separated, after which they are united with suitable mates. The same applies, of course, to those who have been unmarried on earth–whence the rumor of Swedenborg's approaching marriage to Countess Elisabeth Gyllenborg, née Stierncrona.[207]

Ancient Myths and Contemporary Polemics in the Spiritual World

Some of Swedenborg's finest stories from the world of spirits take the form of a visit to the societies that consist of angels and spirits from the various ages. The antique notion of the four ages of the earth was important to him throughout his career as a writer, and now it is combined with the idea of heavenly marriages. What is interesting about these memorabilia, however, is not so much what they have to relate but the way they are staged. The description of his pilgrimage to these parts with an angel as cicerone is in itself fascinating and recalls both *Divina Commedia* and *Pilgrim's Progress*—not to imply that Swedenborg actually read Dante's work, which is hardly likely.

The pilgrimage to the angelic society of the Golden Age leads toward the east through a great desert and a dense forest, where many paths could take him astray and bring him to Tartarus: this is obviously a wanderer who remembers the visit Aeneas paid to Hades. But the angel recognizes the signs: it is necessary to follow a row of olive trees entwined by vines, and soon they arrive at a great plain, dotted by thousands of tents. There live the Lord's warriors from the most ancient times, and their dwellings obviously correspond with earthly habitations from the days when they lived as mortals. In conversations with a married couple, the pilgrims learn of the strict monogamy and absolute chastity of the Golden Age; and Swedenborg is given a pomegranate with seeds of gold as a farewell gift.[208]

The next day they set out for the realm of the Silver Age. On the way, Swedenborg notices many figures carved from wood and stone representing human beings and different animals. He asks his Virgil about them:

> They are figures representing different moral virtues and spiritual truths. The peoples of this epoch had a knowledge of correspondences; and as every man, beast, bird or fish corresponds to some quality, therefore each piece of sculpture represents an aspect of virtue or truth, and several together the virtue or truth itself as a whole. They are what in Egypt were called hieroglyphics.
>
> *De Amore Conjugiali,* n. 76

Here we have a concretization of his recurrent remarks about the origin of the hieroglyphs in the theory of correspondence; and the entire representation of the society of the Silver Age is more marked by the *doctrina correspondentiarum* than usual, which agrees with his view of how the Word has been passed down, as we shall see later. When the pilgrims reach their destination, they first see many horses and chariots, which, however, are quickly transformed into men roaming around and speaking with one another. The angel explains:

> Semblances of horses, chariots and stables, seen at a distance, are appearances of the rational intelligence of the men of this epoch. For by correspondence a horse signifies the understanding of truth; a chariot, doctrine about truth; and stables, instruction. In this world, you know, all things appear according to correspondences.
>
> *De Amore Conjugiali,* n.113

Comparable correspondences have already been described in detail by Swedenborg in his treatise on the white horse of the Revelation of St. John (*De Equo Albo*, 1758), in which he interpreted the classical Pegasus myth according to the same pattern. This time he explains the symbolic import of the houses and temples, wall paintings and statues, even moving pictures of rainbows and other light phenomena, all in the form of a dialogue with an inhabitant in one of the cities of the Silver Age. His farewell gift this time is a bunch of grapes with vine leaves, and these leaves are suddenly transformed into silver—an Ovidian metamorphosis with its symbolical mythical motif adjusted to his own all-embracing correspondence system.

In the same way, he visits the realm of the Copper Age with its many villages of wooden houses; wood signifies natural goodness, which the men of this age possessed. Next in turn is the Iron Age, a kingdom bounded by a thick forest of oak and chestnut, where the guards are in the guise of bears and leopards. The cities are irregularly constructed, the materials wood and brick, the walls hung with idols—a picture of confusion, but not yet of obduracy and hopelessness.

This comes in the fifth epoch, which corresponds to the feet of iron and clay in Nebuchadnezzar's dream of a statue in the Book of Daniel (2:31–45). The people of this age are not able to differentiate between conjugal love and licentiousness, and they are seized with rage and a lust to kill when Swedenborg and his companion try to teach them the truth. But the men of light know how to defend themselves:

> By power given us by the Lord we held up our hands, and fiery serpents, vipers, hydras and dragons appeared from the desert, and rushed in and filled the city, from which the inhabitants fled in terror.
>
> *De Amore Conjugiali,* n.131–132.

Thus we have been told of marital love in its celestial, spiritual, natural, and sensual form in the four ages of the world of classical mythology and Old Testament eschatology; and finally Swedenborg has given us a sense of scortatory love, the evil opposite of marital love. He does this in pedagogically effective scenes, which to him must have met all the requirements of hieroglyphic significance. With the supreme freedom from limitations in time and space that characterizes life after death, he is able to preach his gospel to diverse otherworldly audiences. For example, a number of memorabilia are set in the heavenly Athens, to which Swedenborg is brought by an angel to meet the ancient sages who had exerted such a decisive influence on his life:

> Looking ahead from a considerable prominence I beheld a city, and to one side of it two hills, the one nearer the city lower than the other. He told me, "That city is called Athens, the lower hill, Parnassus, and the higher, Helicon. All are so called because in and about the city dwell ancient wise men of Greece, like Pythagoras, Socrates, Aristippus and Xenophon, with their disciples and neophytes." I

> asked about Plato and Aristotle. "They and their followers," he said, "dwell in another region, for they taught matters of reason which are of the understanding; but these taught morals which are of the life."
>
> *De Amore Conjugiali,* n.218

Newly arrived spirits are frequently summoned to Athens to tell about conditions on earth—as true scholars, the Athenians are eager to learn about everything new. The learned audience listens to shocking echoes of the radical discussion of the day, including intimations of evolution theories, but they are also told of Swedenborg's own activities. That these had met with such a negative reaction on earth astounds the spiritual audience:

> How stupid the minds of men on earth are at this day! If only disciples of Heraclitus and of Democritus were here, who laugh or weep at everything! What laughter and lamenting we should hear!"
>
> *De Amore Conjugiali,* n.256

A third memorabile from the same society dwells at length and with obvious pleasure on the social conditions in heaven; and Swedenborg's dual background as a researcher and a civil servant is clearly reflected in his selection: he makes special mention of courts, libraries, museums, schools, and colleges.[209]

In other tales of heaven and the world of spirits, the narrator enjoys contrasting representatives of various European countries in debates and lectures. A meeting to discuss the question of the origin and force of marital love, attended by delegates from seven nations, has an ending that is as surprising as it is characteristic. The prize goes not to one of the speakers but to an African who has been listening to the proceedings:

> You Christians trace the origin of marital love to the love itself. We Africans trace it to the God of heaven and earth. Is not marital love a chaste, pure and holy love? Are not the angels of heaven in that love? Are not the whole human race and the whole angelic heaven therefrom the fruitage of that love? Can anything so preeminent have any other source than God himself, the Creator and Sustainer of the universe? You Christians trace marital vigor or potency to different

> rational and natural causes. We Africans trace it to man's state of conjunction with the God of the universe.
>
> *De Amore Conjugiali,* n.113

This episode not only gives a vivid summary of Swedenborg's philosophy of love, but is also a typical example of his weakness for Africans, which you may look upon as his personal version of eighteenth-century primitivism; he believed that a more unadulterated relationship to God still existed in native peoples, maybe a trace of the medieval legend of Prester John and his Christian kingdom as will be discussed later on. The construction of the episode is also characteristic. Most of the later memorabilia consist of speeches, debates, and dialogues, with extensive use of rhetorical questions, as in the passage just quoted. Swedenborg himself participates in many of them, particularly toward the end of *De Amore Conjugiali.* In these exchanges, he is not plagued by his severe stammer and inability to lecture as in the real world; instead, he presents his opinions authoritatively, on one occasion by reading long passages from one of his earlier works.

Nevertheless, *De Amore Conjugiali* ends on a note of pain and disappointment. It concludes with a memorabile, in which he is the leading character and which recurs in a corresponding place in *Vera Christiana Religio.* There he tells of still another visit to a society of angels, where sages, thirsty for knowledge, ask for news of the world. In reply, he reports on his own activities as a transmitter of heavenly secrets: that the Word has a spiritual significance which stands out clearly since the revival of the science of correspondence; that men have learned of life after death, of the Last Judgment, of the essence of the Lord, Christ and the Holy Trinity, of the new church. Undeniably, he has been assigned a task more wonderful than anyone else, and he has made tremendous efforts to fulfil it; nonetheless, he is sad and the angels are told the reason:

> Because although the arcana revealed today by the Lord surpass in excellence and worth the knowledge hitherto made known, they are nevertheless considered on earth as of no value.
>
> *Vera Christiana Religio,* n. 848

The angels have the opportunity to establish experimentally the accuracy of Swedenborg's evaluation. On request, they write down the secrets listed on a piece of paper, which is then released from the heavens:

> And behold, in its descent while still in the spiritual world it shone like a star; but as it dropped down into the natural world the light disappeared, and by degrees, as it descended, was veiled in darkness. And when it was let down by the angels into companies where there were learned and accomplished men—from among clergy and laity—a murmur was heard from many of them, in which were these words: "What is this? Is it anything? What does it matter whether we know these things or do not know them? Are they not offspring of the brain?" It seemed as if some of them took the paper and folded, and rolled, and unrolled it with their fingers, to obliterate the writing. And it seemed as if some tore it to pieces, and as if some wanted to trample it under foot. They were withheld by the Lord from that outrage, however, and the angels were commanded to withdraw the paper and guard it.
>
> *De Amore Conjugiali,* n. 533; *Vera Christiana Religio,* n. 848

Swedenborg mentions, finally, that the Lord had revealed the truth about marital love to him, but he doubts that humanity is willing to accept the message. At that point, the angels urge him to write a book and make the revelation known to humankind, a book that the angels will send down from heaven. This obviously refers to the work which the memorabile concludes. Undeniably, a treatise that specifically names the angels as the publishers is a rarity in the history of literature; but after a quarter of a century of daily interaction with spirits and angels, Swedenborg did not hesitate in face of the most provocative consequences. That the wounded pride of the author is reflected here is indubitable and not surprising to the historian: no man wants to speak to deaf ears, particularly if he is convinced of his mission as a prophet.

Conclusion

The survey of Emanuel Swedenborg's interpretation of the secrets of heaven represents merely a collection of examples, but some sense of its unique character may have emerged. One cannot help noticing how he grew ever more confident in his mastery of expression that he had gained from the system of correspondence. In the *Spiritual Diary*, one occasionally has the feeling that Swedenborg was overwhelmed by his experiences and unable to direct them. The published works, however, particularly the later ones, give an increasingly strong impression of literary awareness and deliberation: it is the philosophy at which he has arrived that determines his visions, rather than the contrary. He frees himself from his scientific heritage in various ways. His desire to explain the behavior of the spirits with the aid of the doctrine of form and other psychophysiological theories declines as he comes to understand the biblical authors when they say that one must speak to men in terms that they can comprehend.

This means, of course, that his own memorabilia must be interpreted in the light of the doctrine of correspondence. A distinction in *Arcana Coelestia* between representatives and appearances reveals the need for different categories of symbols, but it is not subsequently exploited explicitly or systematically.[210]

What appears to become increasingly essential for Swedenborg is to communicate a picture of immortal existence as a state of life and activity, of useful work and joy of learning, of love and fellowship; and, to scholars at least, his rather idealized eighteenth-century world appears to be very attractive. The dream of an individual's intrinsic worth and of his or her future life finds strangely concrete expressions in Swedenborg, expressions that are related to his scientific background and also to later trends in the history of ideas and of literature. Before touching on these traditions, however, we must consider Swedenborg's theology, particularly his way of interpreting the Holy Writ.

8

The Lord and His Word

Swedenborg's Hermeneutics and the Allegorical Tradition

It is self-evident that Swedenborg was at home in the world of the Bible at an early stage. Nor is it surprising that he fitted his biblical studies into his scientific research program, as outlined in 1734, in view of its universal aspirations. But what is the background, beyond that stated at the time, of the hermeneutics to which he devoted his enormous energy after 1745?

When we look for an answer to this question, we, unfortunately, discover that no thorough historical investigation of Swedenborg as an exegete has ever been made.[211] This circumstance is highly indicative of theologians' evaluation of Swedenborg's contribution, but it is regrettable, since the lack of such studies makes it more difficult to determine the potential influences of his theological environment. Among his contemporaries, we often find mention of sources of inspiration, most frequently, perhaps, Jacob Boehme and Kabbalahism, which need clarification.

With regard to Boehme, Swedenborg himself had something to say. One of his few Swedish disciples asked the master in a letter for his opinion of Boehme and a certain "L.": the initial probably refers to the well-known English follower of Boehme, William Law (1686–1761). In his reply, Swedenborg denied that he had read them. In reality, he had been forbidden to read theological works before the heavenly secrets were revealed to him, and the reason given is the usual one for his physiological turn of mind: he should be spared from learning heresies that would be difficult to eradicate. When his inner vision was finally opened, he would first have to learn Hebrew and to establish the correspondences of the Word, and for this reason he came to read the Bible repeatedly.[212] This fact is also fully confirmed by the great volumes of indices.

Swedenborg's denial of Boehme's influence is of late date—1767—but such an influence has nevertheless been postulated by many later critics. However, no one has been able to present any proof; if one takes the trouble to make a comparative study, it appears quite clearly that there are no evident similarities, merely such general ones as result from all searches for a spiritual meaning beyond the letter of the Word. The same seems to apply to J. C. Dippel (1673–1734), the radical pietist who paid a dramatic visit to Stockholm at the end of the 1720s, at which time Swedenborg may even have met him: a certain theological influence is hinted at in *Diarium Spirituale*, but cannot apply to the exegesis.[213]

In his letter, Swedenborg stresses what he had said so many times in his writings: that to him the Word was the source of all theology, which implies, among other things, a categorical rejection of all natural theology. If the evaluation of the Word in principle is immutably sublime until the very end, there are, nevertheless, considerable differences in details and traces of an evolution. This does not apply to his conviction of the correspondence character of the Word, but rather to his opinion of its tradition, its levels of significance and kinds of style—always, however, with the reservation that much that appears to be contradictory or new may be due to obscure phrasing and different contexts in the massive theosophical volumes.

As already mentioned, Swedenborg had made a preliminary grouping of biblical texts in different correspondence categories even before his acute spiritual crisis. It is noteworthy, however, that this grouping was never used in later writings. The next time he tried to systematize the levels of significance of the Word—in the preliminary studies for *Arcana Coelestia* published after his death and entitled

Adversaria in Libros Veteris Testamenti (*The Word of the Old Testament Explained*, as they were called in Alfred Acton's translation)—the classification system is different and seems rather to be a return to the preparatory stage of the theory of correspondence. The attempt is made after an explication of Genesis 28:12–15, which deals with Jacob's dream in Betel and the promise that his seed would be disseminated over the earth.[214]

The first level is that of the letter, *sensus externus, historicus seu Literalis*; it thus concerns only the event or the individual explicitly mentioned in the text, or the historical Jacob in the case of the example. The second significance refers to the descendants of the individuals mentioned and their destinies, which, in the example, means the Jewish people: this is known as an inner or higher significance *(sensus internus, superior ac universalis)*. The third level comprises an even wider domain: in the example all those who in any way are in communication with Jacob, that is, the Christian church (*sensus interior, universalior, coelestis et spiritualis)*. Finally, there is a fourth and supreme level, which refers to the Messiah and thus the entire future of world events; this level is labeled *intimus, universalissimus, divinus* and is the plenitude of the truth. In the hypostatical phraseology so typical of Swedenborg in such contexts, he asserts that the preceding significances regard the last-mentioned one as their innermost, their purpose, life, soul, and spiritual light. The Word thus comes to correspond to creation in general and to humankind in particular. For in the innermost depth of all people, there is the soul, which in the shape of the spiritual fluid acts as giver of life even to the outermost tissues; and the four *facultates* from *anima* to the body clearly find their counterparts in the number of meanings in the Word.

Swedenborg thereby defined the model that he followed in his new sphere of activity: it is the psychophysical doctrine of series and degrees from *Oeconomia Regni Animalis*, which he called the premise for the coming universal language; and since the Word in principle constitutes this universal language, this application is a natural consequence. It obviously does not exclude impulses from scientific studies of the Bible, particularly since the practical detailed examination of the texts must have forced him down from the sublime heights of ontology to a more concrete philology. When he formulated the interpretation of Genesis 28:14, his point of departure once again was a Latin translation, but his Hebrew studies were soon to begin.[215]

It is easy to associate between Swedenborg's four levels of significance and the medieval exegetical practice, which was researched most exhaustively by Henri de Lubac in an admirably learned work with the subtitle *Les quatre sens de l'Ecriture* (The Four Meanings of the Word).[216] If a connection could be proven, it would undoubtedly provide a more solid foundation than is offered by the concept of "allegorical biblical interpretation." This phrase is constantly used in the Swedenborgian literature, but it is ambiguous to the point of meaninglessness and comprises elements that Swedenborg rejected outright.

The allegorical method of reading texts is encountered as far back as in the Greek Homeric interpretations of the sixth century BCE; by the time of Jesus' birth, it had reached a high degree of systematization in the Stoics and the Neo-Pythagoreans. The rabbinical tradition also included this allegorical method, although clearly distinguished from the Greek. Where the Greeks looked for philosophical propositions in mystical disguise, the rabbis wanted to achieve as complex an interpretation as possible. However, the Jews of the Diaspora, particularly Philo Judaeus, adhered completely to the Greek allegoristic method, and they passed it on to Christian hermeneutics.

There is no need to question Swedenborg's general understanding of these things, which is proven both by his own studies and by his long relationship with Erik Benzelius, a prominent specialist on Philo. How familiar Swedenborg was with the rabbinic tradition and the Kabbalah is uncertain, but one can find numerous examples of an extremely negative attitude on his part: almost concurrently with the explication of Genesis 28:14, he anathematized all philosophy (with the exception of his own works, to be sure) in *Diarium Spirituale*, and included in philosophy "fables and silly stories, especially such as have formerly and do still distinguish the Rabbinical writers, which are innumerable."[217] Statements of this nature should make us extremely cautious about attributing any influence on Swedenborg by the Kabbalah, for example, which has sometimes been done.

Through his extensive reading, which is documented in his excerpts and in his printed treatises, Swedenborg became familiar with exegetic principles beyond those he had learned in his youth and through "incidental learning" as a member of a large clerical family. At the same time, it should be emphasized that no traces remain of systematic studies of the nature of the philosophical researches in cod. 36–110. His knowledge of as tremendously an influential writer as St. Augustine was probably considerable and may even have included the hermeneutic discussions in *De Doctrina Christiana.* Augustine was of decisive importance to the success of the Alexandrine allegorists.[218] A part of this tradition was a legacy from Origen, who established similar correspondences between the Word and humanity to those we find in Swedenborg. According to Origen (and St Paul!), the human being consists of body, soul, and spirit, and the Word that refers to our salvation must have three corresponding levels of significance—a literal, a moral, and a mystical: "Like man, it has a body, a soul and a spirit; whence the terms of a corporeal sense for history, a psychical sense for morality, and a spiritual sense for allegory (or anagogy)."[219]

Other scholars replaced these three levels by four, but de Lubac's study makes it clear that this did not represent any basic change, since the same writer sometimes works with four and sometimes with three significances. Nicholas of Lyra (c. 1270–1349) wrote a widely disseminated mnemonic rhyme, in which the four levels are named, but its underlying reality is far more complicated than the simple Latin in which it is expressed

> Littera gesta docet, quid credas allegoria,
> Moralis quid agas, quo tendas anagogia.[220]

These four significances form two groups, called *sensus litteralis vel historicus* and *sensus mysticus seu spiritualis*, respectively. According to Nicholas, the latter is thus tripartite, and the difference between the groups is that the import of the first one is *per voces*, according to the words, while the meaning of the second is *per ipsas res*, according to the things themselves. The latter is divided, as the verse tells us, into *sensus allegoricus, moralis vel tropologicus*, and *anagogicus.* The first of these is involved if that which is said shall be referred to what we should believe according to the new law. It is thus identical with what others call figural or typological significance, an event in the Old Testament that prophesies something in the New Testament, or, to use Swedenborg's earlier terminology, a case of *correspondentia typica.* The second spells out a moral instruction from the biblical text, and the third refers to what we can hope for in heaven. The name *Jerusalem* is the example most frequently used in these contexts. Literally, of course, it means the city in Palestine, allegorically it represents the Christian church, tropologically the soul of man, and anagogically the heavenly city in the Revelation of St. John the Divine.

If we stop for a second to draw a comparison with Swedenborg's conceptual arsenal, we see both similarities and differences. In *Clavis Hieroglyphica* we find four categories of correspondence, all of which, however, clearly belong under Nicholas' *sensus mysticus.* Furthermore, only one of them corresponds in any sense to the traditional, namely, *correspondentia typica.* In Swedenborg's version, however, this typological significance also embraces the anagogical. The classification in *Adversaria* is actually more reminiscent of the medieval practice. There we find *sensus litteralis vel historicus* and three inner levels of significance. None of them, however, adheres fully to Nicholas' definitions, but Swedenborg appears to have roamed very freely within the traditional framework, regardless of how familiar he was with its details. The pattern corresponds best with Origen's, and the correspondence is even greater at a later stage, when Swedenborg included *animus* in the body and thereby reduced the number of *facultates* to three. We shall return to this point shortly.

Swedenborg was not content with abstract speculations of this nature. His ruminations on the problems of interpretation can be illustrated by comparing the three different interpretations of the first chapter of Genesis that he produced in the years between 1745 and 1749. In the one immediately succeeding the drama of creation, *De Cultu et Amore Dei*, the principal aspect is still cosmogonical, and the "allegorical" elements are confined to his making the days of the creation represent very long periods of time, which, of course, is completely in accordance with an ancient Christian tradition.[221]

The next version, which introduces the real *Adversaria*, does, admittedly, take up the cosmogonical aspects; but the point of view has now been altered to what a medieval exegete would probably call anagogical. Swedenborg claims that he now wants to examine the Word to learn about *civitas Dei*, the ultimate purpose of the creation. This means that the emphasis in the first chapter will be on man in his state of integrity, the first reflection of the Kingdom of God in nature:

> He was born in the most perfect order, that is, after the likeness of God; thus the last was represented in the first. Wherefore nowhere else in the Old Testament, except in the first-born beings before the fall, can we contemplate an effigy of that kingdom; namely, that in it will exist a like order, but still more perfect because existing in an entire society which will form one body and will live with one spirit; and thus, by means of the one only Love by whom that order commences and in whom it ends, will come a most perfect image of God.[222]

In *Arcana Coelestia,* we find a radically consistent application of this symbolism. Now Swedenborg ignores the literal significance and gives only the inner meaning, which, admittedly, is frequently differentiated, but without the rigid systematization of the previous draft. It is obvious that his greater experience with practical hermeneutics and his study of Hebrew had partly changed the nature of the problem. One illustration of this is that, as early as in his commentary on the first chapter, he tries to give a brief survey of what he calls the different styles of the Word, which is intended to clarify his evaluation of the text *in concretu* and of the historical background.

Arcana Coelestia

According to *Arcana Coelestia* n. 66, there are four types of style in the Word. The first is that of the most ancient church and expresses an exclusively spiritual meaning:

> their mode of expression was such that when they mentioned terrestrial and worldly things they thought of the spiritual and celestial things which these represented. They therefore only expressed themselves by representatives, but also formed these into a kind of historical series, in order to give them more life; and this was to them delightful in the very highest degree.
>
> *Arcana Coelestia, n. 61*

It appears from subsequent passages that the difference between this church and the next one—the one that came after the Deluge—was the exact counterpart of the difference between the celestial and the spiritual. The people of the most ancient church spoke not in words but through direct transference of thoughts and facial expressions. Where they experienced a direct inner sense of the true and the good, the next era instead had "another kind of dictate; which may be called conscience" *(Arcana Coelestia,* n. 607).

The differences are also reflected in style. Until the days of Abraham, the Word was written in reproductions of the most ancient language, which Moses inherited from his ancestors, but the remaining books of Moses and the books of Joshua, Judges, Samuel, and Kings are written in the second style, which Swedenborg calls the historical. In these books, historical material is presented in the literal sense, although there is also an inner meaning, which he promises to reveal later. The third style, that of the prophets, sprang, according to Swedenborg, from the original church: the difference is that its character is not so consistent or so "historical" but more disparate, which makes it almost incomprehensible unless one has the key to the inner meaning. Finally, Swedenborg distinguishes a fourth level of style in the Psalms, "halfway between prophetic and everyday speech."

This means, therefore, a rough classification in "mythical," historical, prophetic, and poetic parts, which in itself is not particularly original. Nor is it surprising that Swedenborg was later compelled to make certain adjustments and to distinguish intermediate forms. What is original is his way of combining these stylistic observations with the theory of correspondence. The pervading idea is that the members of the most ancient church lived a symbolical existence characterized by correspondences, but that their unconscious and intuitive spirituality was lost after the Deluge. This spirituality was then replaced by a conscious but not fully realized symbolism, a *theory* of correspondence but no longer a *life* of correspondence; and this doctrine found different expressions: in the Jewish church in liturgical rites and symbols and in the Holy Writ, in Egypt in hieroglyphics, in Greece in mythology, etc. Swedenborg considers that the concealment of the inner significance of the Word under a "historical" mantle is justified pedagogically: it shall make such thrilling reading that even children will be attracted to it.

Naturally enough, there are a number of other distinctions in *Arcana Colestia*, including a classification of the meaning of letters in three subsections, *historicus*, *propheticus*, and *doctrinalis*, but no really decisive changes in the textual-historical pattern are to be found in this work. Since Swedenborg here is trying to disclose the spiritual significance, his expressions of the literal meaning are often negative. This appears most clearly in a paragraph toward the end of the great work, where he states categorically which books shall be regarded as the Word, namely, only those that have an inner meaning. All the books of the Old Testament are included, with the exception of Ruth, Chronicles, Ezra, Nahum, Esther, Job, Proverbs, Ecclesiastes, and the Song of Solomon, while the New Testament is represented only by the Holy Gospels and the Revelation of St. John the Divine *(Arcana Coelestia*, n. 10325). This obviously is highly restrictive; and it is tempting to regard it as self-doubt on the part of the author when faced with the tremendous task of revealing the spiritual significance.

The Doctrine of Correspondence in the Hermeneutic Function

After his analysis of Genesis and Exodus in *Arcana Coelestia*, Swedenborg presented no systematic study of any other Bible book than the Revelation of St. John the Divine, but obviously all the theosophical treatises contain innumerable textual interpretations. He also published a special small tract on the Bible in the 1760s, many of the doctrines of which recur in *Vera Christiana Religio* and thus represent his definitive standpoint.[223]

Doctrine of the New Jerusalem concerning the Sacred Scriptures, the imposing title of the small tract, is introduced by an argument that Swedenborg had already used on many occasions, namely, that many people do not grasp the sublimity of the Word because its style is so simple and so unlike the flowery mode of expression of the day: one of many attacks on rhetorical ideals of style. In the Word are concealed the most profound secrets, accessible to those who are able to penetrate beyond the letter. From the love of the Lord springs the celestial, which is also the divine good, and from his wisdom the spiritual, which is the divine truth: these two are united in their utmost radiation, the natural. For the Word to be perfect, it must correspond with these three degrees, and the literal meaning thus contains two higher significances, just as an effect contains both its active cause and its final cause.

Clearly this must be a matter of discrete degrees, which, unlike continuous degrees, do not successively merge with one another, but can only be united through correspondence. The Word is thus written with correspondences, and here Swedenborg presents his first survey of the history of the science of correspondence. In general, we are familiar with it from *Arcana Coelestia*, but Swedenborg claims that he has now been taught that the Enoch mentioned in Genesis 5:21–24 had collected these *disiecta membra* of ancient wisdom and passed them on to posterity. Enoch was Noah's great-grandfather, and his descendants spread the tradition throughout the

Orient and from there to Greece, where it was transformed into fables.[224]

This historical vision seems to agree relatively well with the opinion of the dissemination of the art of writing current at the time. During the course of the centuries, however, the *repraesentativa* of the churches, which, of course, were correspondences, were transformed into various kinds of idols, whereupon providence consigned the science of correspondence to oblivion: this is how Swedenborg presents what he had understood from Wolff, thirty years previously, about the metamorphosis of the hieroglyphs into idolatry.

The lost science was not revived until Swedenborg's day, and he gives three explanations for this: (1) the first Christians were too unlettered to understand it; (2) the popes later withheld the Bible from humankind; and (3) while the Reformation admittedly returned the Bible to the laity, at the same time, a theologically unfortunate distinction was made between faith and way of life—Swedenborg's recurring accusation of solifidianism—and this carries a risk of the correspondences being referred solely to faith and not to love also.[225]

Once Swedenborg had revived the science of correspondence, the way was opened to inconceivable spiritual treasures. But the premise is the literal meaning, for which he now expresses great respect. Without the letter, the Word would be like a body without skin or bones, like a brain without membranes and skull, like a heart and lungs without a thorax. When a person reads the letter of the Word, a spiritual angel brings forth the spiritual significance and a heavenly angel the celestial meaning. The latter cannot be visualized, however, since it is not accessible to the intelligence but acts on the will alone. That it exists is proved, for example, by the Bible's predilection for parallelisms, which Swedenborg regarded even in *Arcana Coelestia* as a linguistic correspondence to the marriage between the true and the good, but it can also be experienced as a vitalizing force.[226]

The connection with heaven is opened by this Word, which also came into existence for the inhabitants of heaven and the spiritual world. It applied to earlier churches; but after the incarnation of the Lord, which, as we have learned from the Gospel according to St. John, made him the Word in the world of natural things also, the church no longer occupied a purely representative position, but possessed the full truth as well. Since the Word has always existed somewhere on Earth, the communication with heaven was secured even for those who themselves did not possess it, except in distorted reflections, as, for example, the Greeks and, later, the Mohammedans.[227]

This interpretation of history is complemented in a strange fashion in some of Swedenborg's last works. In his second great exegetic work, *Apocalypsis Revelata* (1766), he reports that he had learned in conversations with Tartar angels that the so-called *Verbum vetustum*, a pre-Israelitic Writ, was still known to the peoples of Great Tartary. It was said to consist solely of correspondences, and it embraced the older traditions referred to in various parts of the Bible: *liber Jaschar, Bella Jehovae*, and *Enuntiata* in Swedenborg's Latin translation.[228] The same information is repeated in *De Amore Conjugiali* in connection with the visit to the angels of the Age of Copper described earlier: these angels are alleged to possess the Asiatic Word, which is now preserved in Magna Tartaria. It also recurs with minor variations in the last work, *Vera Christiana Religio*.[229]

One may wonder about the origin of these strange notions, which Swedenborg himself explicitly attributed to a new revelation. Tartary and its inhabitants belong to the legendary concepts in European history, and Swedenborg's contemporaries appear to have had very vague notions of its geographical position and its culture. Great Tartary, as distinguished from the more westerly areas, which sometimes were called Little Tartary or Crimean Tartary, was believed to lie somewhere in the great table lands of Central Asia.[230] The border between it and China was also unclear; but in *Apocalypsis Revelata*, Swedenborg suggests with the scientist's characteristic desire for empiric certitude, that enquiries in China should confirm what the angels had said about the remarkable Tartars.

At least a part of the explanation of Swedenborg's claim might be found in the medieval legend, according to which the so-called kingdom of Prester John was located in the realm of the Tartars. Beginning in the middle of the twelfth century, there was a rumor in Europe of a

Christian prince in Central Asia; letters were even received from a sovereign who called himself Prester John. When Marco Polo and other adventurous travelers in Asia in the thirteenth century reached Tartar territory, they soon discovered that no Christian king was there; but this did not put an end to the legend. Instead, John's alleged kingdom was transferred first to India and later to Africa. As late as the seventeenth century, the official name of Abyssinia was *Regnum Presbyteri Johannis*, the realm of Prester John, and the legend survived in other forms until the middle of the nineteenth century.

It is difficult to establish how Swedenborg came into contact with this myth, which probably had its origin in lack of knowledge of the history of Christian mission. In any event, his claims are strange expressions of the strong interest in Asia, particularly Chinese culture, which the reports of the Jesuit missionaries aroused in cultivated Europeans, not least in Leibniz.[231] Anders Hallengren has recently pointed out that the Swedish prisoners of war, who spent many years in Siberia after the battle of Poltava, might also be an interesting connection, particularly since Swedenborg's cousin Peter Schönström and his friend Philip Johan von Strahlenberg were among the captured officers, who devoted themselves to mapping and making other researches about Central Asia: Strahlenberg even identified the legendary Prester John with Dalai Lama.[232]

In Swedenborg's own development, these and other similar reports represent the last link in a larger tendency to fit his exegesis into a historical tradition. Swedenborg never claimed to have invented his hermeneutic principles, but constantly stresses that he had revived a form of knowledge originally regarded as the science of all sciences. Obviously, his thesis would receive the best support conceivable if he could find literary testimony more ancient than the Bible and written in the language of correspondences. That Swedenborg's dream was fulfilled by his celestial teachers is only too understandable to readers who have some experience of hypothetical argumentation. That these teachers located *Verbum vetustum* in inaccessible Great Tartary is evidence that their wisdom was as great as their love was fervent.

The Lord and His World

Swedenborg's theology is based solely on the Word and the interpretation of its spiritual import; this he declared himself, and it is patent to any reader. Even though it may sound paradoxical, it is possible to reverse this statement. With the sublime freedom to understand the inner meaning of the Word, which the doctrine of correspondence yielded him, the interpretation was in reality determined by the intellectual system he had already developed. The basic ideas in his theology had been worked out during his scientific period, and much in his later evolution consists of explicit definitions of what previously had been tentative thoughts. At the same time, his thinking after 1745 was to some extent changed and simplified.

With regard to his idea of God, we learned from the discussion of *De Infinito* (1734) that Swedenborg was extremely anxious to maintain a sharp and clear boundary between infinity and the finite state: he issued a special warning against false mathematical analogies beyond this boundary and against identifying God with nature. When he expounded his dogmatic universal theology in his last work, *Vera Christiana Religio*, we find the same strict transcendence in his idea of God. God's being is the substance and the form itself and cannot, therefore, be described with the aid of lower substances and forms, which are merely obscure images of this supreme being. It is a mystery beyond all human comprehension; nor can the angels grasp the being of God, created and thus finite beings as they are.[233]

If God's *esse* (being) is inaccessible, it is, however, possible to understand something of his *essentia* (essence) with the help of the two infinite attributes that are its foremost characteristics, love and wisdom. Swedenborg regards these as the source of life, the spiritual sun—we recall from the celestial scenes in the memorabilia that heaven is constantly irradiated by a sun in the east, which thus represents the Lord in the form of the light of wisdom and the warmth of love.

However, this spiritual sun, which is the prototype of nature's inanimate source of light and heat, is not the only manifestation of the

Lord, nor even the most important one. His first manifestation was when he pronounced the Word, as related in the first chapter of the Gospel according to St. John, and thereby heaven was given human form.[234] The Word is also for the angels in heaven, which are shaped like human beings, and we know from the Old Testament that the Lord revealed himself in human guise even before his incarnation. But the human race grew so remote from him through the centuries that a more radical manifestation was needed to restore contact, and finally the Lord allowed himself to be born into the human world through a mortal woman.[235]

As a theologian, Swedenborg thus rejects both the concept of Christ as the Son of God and the doctrine of the Trinity. This is a more radical approach than in *De Infinito*, where he regarded the Son as the connecting link between the infinite and the finite, but it does not imply any disparagement of the incarnation as such. It was God's intention to defeat the power of hell and to establish a new church, but his hands were tied by the eternal order which God himself had created. Therefore, God must act through humankind. By first assuming human form and then successfully resisting all human temptations, he allowed the human species to become godlike, so that the new covenant could be established along completely new lines. This was the work of salvation, not the death on the cross—that God became human and humans thereby God is the imperishable miracle.

It is possible that Swedenborg was familiar with the kabbalic notion of Adam Kadmon, the original man, when he conceived his vision of the God-world relationship, or that he may have been inspired by similar ideas in other esoteric traditions. Nevertheless, the simplest explanation is that, through the doctrine of correspondence, he expanded the psychophysiological thinking of his scientific period. For, in principle, *anima* (the soul), which forms its body in the egg, is just as incomprehensible as the *esse* of the Lord, and the construction of the organism, like the divine act of creation, means that the spiritual takes on material shape. The difference is, of course, that God created the universe from himself, while *anima* is introduced in a pre-existing material in the egg. For the rest, both activities are incorporated in the system of series and degrees, which is now extended to include the divine.

While Swedenborg in *Oeconomia Regni Animalis* assumed a simple first substance outside all series, the divine *esse* now becomes the highest series, in which everything created existed potentially before the first manifestation of the Lord.

The number of degrees in each series is now set at three. These are, of course, discrete degrees, united through correspondence, and they are related to one another as end, cause, and effect. The end first generates the cause and then the effect in order to realize itself, and this triad is repeated at all the levels of creation. Each effect becomes an end in a new series at a lower level all the way down to the lowest matter, and the entire universal chain is held together with the help of correspondences. Only humans are conscious of this constant creative process, and the primary end of creation can only be achieved through the coming into being of humans, who can bind together the ultimate and the first by acknowledging the universal teleology and by voluntarily associating themselves with the divine.[236]

Free will is obviously a cardinal point in Swedenborg's thinking, as is true of Leibniz and Malebranche, the "philosophers of order," who were his most likely sources of inspiration. The basic concept of a transcendent God beyond all human determinations, from whom emanates a cosmos which is irradiated by his force all the way down to the most obscure reflections in matter, cannot help but remind us of Neo-Platonism, and it is probable that Swedenborg himself made the same associations. This does not mean that he consciously aligned himself with Plotinus, but it qualifies him as belonging to the tradition often encountered under the name of *philosophia perennis*, timeless philosophy.

The term suggests an intellectual current that maintains its identity despite transitory ripples on the surface: in the Western world, the term usually implies a Platonized interpretation of Christianity with pronounced mystical elements, and its great proponents include Origen, Augustine, Dionysius the Areopagite, Eckhart, and many others. Among these, Swedenborg is known to have been deeply interested in

Augustine, which is natural in view of the attitude of contemporary masters, e.g., Leibniz and Malebranche, two modern names in the tradition of *philosophia perennis.*

We find serious differences of opinion between Plotinus and Swedenborg on the subject of freedom of the will and the origin of evil, which exemplify the foregoing remarks. To Swedenborg, no matter exists beyond that which is the final link in the golden chain and thus allied with God, and he therefore cannot identify matter with evil, as Plotinus did. In a memorabile, in *De Amore Conjugiali,* Martin Lamm discerned an attack on Plotinus' theory that light in its uttermost emanation is lost in darkness; that is another way of saying that good is absorbed by evil.[237] To Swedenborg it is quite clear that the only source of evil is humanity's sin, which means that we have diverted our love from God and thus from its ultimate end to our own ego. However, sin did not come into the world as the result of a single act but through a successive rejection of God, a progressive love of self, a counterpart of the gradual loss of the science of correspondence.

In this way, Swedenborg came to reject the notion of a revolt in heaven and a personal devil, ideas he had formulated in *De Cultu et Amore Dei*; and as a theologian, he gives humankind the entire responsibility for its own fate and thereby for that of the created world also. But, like all other advocates of theodicy, he was also compelled to take a stand on the question of the relationship of God to evil, once it had arisen. The association with Malebranche's idea of order is revealed most clearly in his answer.

God's almightiness cannot be compared with that of a mortal tyrant, who can obey any personal whim; while God, in his love and wisdom, is order itself and can only do good. People have been given free will and must, by virtue of the divine order, be allowed to use it freely. If we turn away from him, God cannot force us to change our attitude, but he can approach us with the warmth of love and light of wisdom to the extent that we want to be near him: and God has provided all people with good spirits, who will help them in the struggle against evil and will counterbalance the influence of the emissaries from hell.[238]

Swedenborg does not reckon with original sin in the Lutheran sense; he believed that hell came into existence as a result of humankind's having sinned almost throughout its history and having thus inherited an inclination for sin, which is handed down from one generation to the next. Like other reformers, Swedenborg also preaches the gospel of regeneration, but his version bears the stamp of the intellectual self-analyst. What he demands of people is that they search their souls for their innermost motives, as he himself did in the critical years in the mid-1740s, and that they then consciously and actively combat all the evil revealed by their self-analysis.

To Swedenborg, a regeneration cannot take place through faith alone, as in Lutheran orthodoxy, nor as an immediate transformation of personality resembling the conversions of Methodism; like the Fall, it is a continuous process and expresses itself in labors of love vis-à-vis God and one's neighbor: it also transcends death and comes to fruition in the intermediate stage immediately after the body has perished. That which is regenerated is the inner person, and for those who, with the help of the Lord, have succeeded in accomplishing the act of rebirth, the fullness of God's love and wisdom pours into this inner person, and he or she lives under the direct guidance of *anima*, in the same way as the first people lived. Swedenborg had spoken nostalgically of the bliss of the first humans ever since the preface to *Principia* in 1734; as an aging theologian, he at last was privileged to experience it in his own life.

But his lot was even more blessed. He was chosen to show his brothers and sisters the way to a realization of God's great purpose with the creation. He was chosen to found the new church mentioned by St. John the Divine as New Jerusalem. In his interpretation of the Word, the promise of the second coming of the Messiah was fulfilled. In the new church, discord between faith and science would no longer plague a humankind whose frantic search for knowledge was so familiar to Swedenborg. A memorabile in *Vera Christiana Religio* describes the new church as a temple with a pearly door and walls of crystal. Over the door is the inscription *Nunc licet*, the interpretation of which sounds like a strange variation of the many proclamations of the

Enlightenment to the effect that humanity had achieved maturity: it was now permissible to penetrate the secrets of faith with the intellect.[239]

The way was opened by Swedenborg's exegesis. Few intellectual constructions and systems of faith offer a nobler dream of mankind and a more breathtaking vision of man's fate than Swedenborg's New Jerusalem.

9

New Jerusalem in the World

One afternoon in August 1764, Carl Christopher Gjörwell, head of the Royal Library, visited Swedenborg in his house on Hornsgatan in Stockholm. The visitor found him in the garden tending his flowers and was invited to take a stroll and enjoy the splendor of late summer. Gjörwell's errand was to request Swedenborg to present his most recent works to the Royal Library. This request was received very graciously: "Most willingly," Swedenborg answered and added, "Besides, I had intended to send them there, as my purpose in publishing them has been to make them known, and to place them in the hands of intelligent people." As they walked, the kindly and sprightly old man gave the librarian a detailed report on the fundamentals of his theological system interspersed with entertaining glimpses of the spiritual world.[240] Later, in the August issue of his periodical, *Svenska Mercurius,* Gjörwell reported that all Swedenborg's theological works had been received by the Royal Library as a gift from the author.[241]

This idyllic scene is characteristic not only of Swedenborg's friendliness and his sunny nature, documented by many of his contemporaries, but also of the manner in which his ideas were disseminated. Gjörwell's visit occurred a few years after the learned of Stockholm had become aware of the celebrity in their midst, and the librarian himself had spread the word in earlier reports in his publication. In a letter from the beginning of the 1760s, Swedenborg's former assessor colleague Daniel Tilas wrote that "the whole town" had recently buzzed with talk about Swedenborg's mystical talents, of which nobody had ever heard; due to the theological censorship in Sweden, the great theosophical volumes had been published abroad anonymously.[242] It was at about the same time that Swedenborg began to attract attention abroad, particularly in Germany through Johann August Ernesti's extremely negative criticism and Friedrich Christoph Oetinger's positive and well-informed reviews.[243]

Swedenborg never played the role of a sectarian preacher but relied wholly on his writings, which, of course, he hoped would reach as many qualified readers as possible. Despite the prevailing censorship, he seems to have had no problem in giving away copies of them in Sweden until the last few years, when the consistory in Gothenburg held a minor heresy trial of his first real disciples in the country. It was by no means only the Royal Library that received his books, but a long series of the temporal and spiritual lords of the realm, with whom Swedenborg maintained close relationships. Perhaps the foremost of them all, Count Anders Johan von Höpken, one of the most brilliant cultural personalities of eighteenth-century Sweden, was fascinated by Swedenborg's theological philosophy and impressed by his intellectual gifts, though doubtful of the value of the descriptions of the life of the spirits.[244]

In attempting to sketch Swedenborg's importance in the history of ideas, it is, of course, essential to differentiate between direct influences, conveyed by various kinds of Swedenborgian societies and congregations, and indirect effects via individual literary works. Obviously, the two sometimes coincide; but, in general, the purpose of the many New Church congregations was to make converts within or without the existing Christian churches, while Swedenborg's influence on *belles lettres* is in no sense confined to those who wholeheartedly embraced his faith.

The earliest effects of his authorship in Sweden can be divided roughly into two groups, which can also be defined geographically.

First, we have theological influences in the limited sense of the term, which were concentrated in the dioceses of Gothenburg and Skara in the western part of Sweden and which began to be felt at the end of the 1760s. It is true that this resulted in considerable unrest, even a heresy trial in Gothenburg; but, in general, Swedenborg's rationalistic theology could be accommodated to the accepted doctrines of the Church of Sweden; and it is remarkable how many priests in western Sweden were in sympathy with Swedenborg's teaching in the last decades of the eighteenth century and the nineteenth.[245]

The other group had its center in Stockholm with strong ties to the tolerant court of King Gustavus III. As has been demonstrated in recent Swedish scholarship, it was of another type, occult and syncretistic in the sense that Swedenborg's theories were found to provide a rational explanation of all kinds of mysticism, from alchemy to animal magnetism; indeed, the king was the last European sovereign to keep his own court alchemist.[246] And it was in Stockholm in the middle of the 1780s that the first Swedish society for the translation and publication of Swedenborg's works was founded, the so-called Exegetic-Philantropic Society; it managed to get a few small sections into print before the authorities put their foot down.

The ban was effected on the initiative of the church, but strangely enough it was supported by the radical advocates of the Enlightenment. This unholy alliance can be explained by the occult aspect of the society, which made it the target for journalistic attacks, the sharpest and most skillfully worded of which was written by the brilliant poet Johan Henrik Kellgren.[247] In the eyes of the educated, Swedenborg came to be a symbol of folly and befuddled thinking, a stigma that has still not completely disappeared, even though the generation of writers that succeeded him, the young romanticists of the first decades of the nineteenth century, did a great deal to correct Kellgren's profoundly unfair picture.

However, it was outside Sweden's borders that Swedenborg's influence was greatest: this was natural in the light of his own citizenship in the international republic of the learned. The reactions in the nearest major cultural area, Germany, appeared at about the same time as in

Sweden, in the 1760s; and we have mentioned on several occasions the Swabian theosophist F.C. Oetinger as the first example of a perceptive and fascinated reader of Swedenborg. But Oetinger did not determine the German attitude, even though his support resulted in the publication of collections of Swedenborg documents at a very early stage and in the translation of his works.

It was a much more powerful spirit that influenced the German image of Swedenborg, namely, Immanuel Kant. He must also be given much of the blame for the distortion of Swedenborg's work that arises when the entire emphasis is placed on the conversations with spirits. Kant was one of the first readers of *Arcana Coelestia*, and the experience made him extremely disappointed in the author, to whom he was favorably predisposed. It was in this mood that he wrote his biting attack, the famous *Träume eines Geistersehers* (Dreams of a Spirit-seer, 1766).[248] The tract introduced Kant's accounting with the principles of traditional metaphysics which culminated in *Critique of Pure Reason* fifteen years later. Swedenborg's admirers can thus find some comfort in the circumstance that his negative influence had extraordinary effects.

But Swedenborg's importance to German intellectual evolution is not entirely negative, even though Kant's scornful pamphlet made his admirers cautious and reticent. In a review of Lavater's *Aussichten in die Ewigkeit* (Perspectives of Eternity) in 1773, the young Goethe advised the later-so-famous physiognomist to study Swedenborg, superfluous advice, indeed, since Lavater had been in correspondence with the author of *Arcana Coelestia*, but it does reflect the reviewer's positive attitude.[249] Despite Kant's one-sided and strangely ill-informed condemnation, Swedenborg became an active force in German tradition, as one can see in Schelling's, Görres's, and other romantic thinkers' interest in his theosophy. In G.H. von Schubert's important *Symbolik des Traumes* (Symbolism of Dreams,1814) exhaustive quotations from *Arcana Coelestia* serve as illustration of the particular character of the language of dreams.[250]

Sometimes Swedenborg was even included in the German tradition. In his widely read and discussed book, *Rembrandt als Erzieher* (Rembrandt as Educator, 1890), Julius Langbehn wrote that

Swedenborg's doctrine of the unity of organisms corresponds "to the thinking of every truly German spirit"—a greater tribute could scarcely have been paid in such a culturally chauvinistic context.[251] Even in recent years, one may stumble over such allusions to Swedenborg that prove him to be an obvious reference for literate people: one rather amusing example is offered by Martin Walser in his novel *Ein fliehendes Pferd* (A Runaway Horse, 1978).[252]

The Exegetic-Philantropic Society in Stockholm had active relations with a number of occult counterparts in Germany, but its closest prototypes were the Swedenborgian societies founded in England in 1782 and 1783; many personal relationships were established with members, and there were certain transactions with manuscripts and letters in Swedenborg's handwriting.[253] The first independent Swedenborg congregation, the origin of the worldwide New Church, was founded in London at the end of the 1780s, and it is primarily due to the efforts of the English and American societies that Swedenborg's works were translated and his manuscripts published—a magnificent effort, not least when one recalls that the membership has probably never exceeded 20,000.[254]

In the same decade, Swedenborg was introduced in the United States of America. On June 2, 1784, the following advertisement appeared in the *Pennsylvania Gazette*, as Marguerite Beck Block tells us in her book on Swedenborgianism in America:

> For the Sentimentalists. A Discourse on the extraordinary Science of Celestial and Terrestrial Connections and Correspondences, recently revived by the late honorable and learned Emanuel Swedenborg, will be delivered by Mr. James Glen, an humble Pupil and Follower of the said Swedenborg's at 8 o'clock on the evening of Saturday the 5th of June 1784, at Bell's Book-Store, near St. Paul's Church, in Third St., Philadelphia.[255]

The contributions of these enthusiasts led to Swedenborg's becoming a part of the cultural heritage of England and the United States to—at least sometimes—a far greater extent than in his homeland, and a very long list of prominent readers and disciples could be cited.

William Blake was one of the signatories of the conference resolutions in the newly founded London congregation in 1789, and he dedicated mind and heart to his research on Swedenborg during the next few years. But his revolutionary spirit could not subordinate itself to any system, neither Swedenborg's nor, even less, Swedenborg's disciples' interpretations of it.

Instead, the most tangible reminiscences of Swedenborg in Blake's poetry are in the form of parody and satire. The very title of his remarkable *The Marriage of Heaven and Hell* (1790) is a critical allusion to *De Coelo* (1758)—known as *Heaven and Hell* in English—which expresses the poet's aversion to Swedenborg's strict distinction between good and evil. The disposition is also a parody of the master: a Memorable Fancy concludes every chapter, comparable to Swedenborg's memorabilia.

Despite these defiant gestures, Blake received decisive impulses from Swedenborg, not least concerning the interpretation of the Bible; and his reading of newly translated works such as *Divine Love and Wisdom* and *Divine Providence*—a temperamental reading with pen in hand—aroused his need for a personal stand on problems of decisive importance.[256]

William Blake is admittedly one of the most profoundly original geniuses in the history of literature, but his ambivalent attitude to Swedenborg can nevertheless be regarded as typical of the reactions of the poets. It is hard to find any great writer who accepted Swedenborg without reservation or who involved himself in any New Church congregation. Therefore, it is frequently difficult to determine Swedenborg's influence. Henry James Sr. overcame a severe mental crisis in the middle of the 1840s—a *vastatio* in Swedenborg's terminology—as a result of being advised to read Swedenborg; the first works he read were those studied so thoroughly by Blake. This experience made James Sr. a faithful though independent disciple for the rest of his life. He published anonymously *The Secret of Swedenborg* and had contacts with Swedenborgian circles. James Sr. never joined the New Church, but evolved a personal interpretation of Swedenborg without proselytic aspirations. In the words of Henry James Jr., "The temple of Swedenborg stood in the centre of our family life."

What appears to have been of greatest concern in this temple was the morally activist aspect of the master's teaching, the experience of the creation as an unceasing progression toward the kingdom of God, the belief that human beings can help to hasten the achievement of this goal by conquering egoism and attachments to the good things of this world. Neither Henry nor William James could share their father's optimism or accept his visionary master, but there is no doubt that they received indelible impressions from the temple in the center of their family. This is particularly noticeable in the older brother William, who himself experienced a *vastatio* resembling his father's and wrote of it in his great work on the varieties of religious experience.[257]

One of the James family's closest friends was the great American author who drew what is perhaps the most important and most influential of the many Swedenborg portraits in literature. When Ralph Waldo Emerson published his famous collection of essays, *Representative Men* (1850), he chose Swedenborg to represent mysticism.

This essay has become such a standard work for the profane evaluation of Swedenborg that to read it is constantly to meet old acquaintances. Emerson wanted to give an overall picture, in which the scientific period is attributed as much significance as the theosophical, and he was fully aware of the close connection between them. He also reproduced the intellectual milieu in a way that admittedly is not correct in every detail but that, nevertheless, reduces Swedenborg's originality to reasonable proportions. And he treats the visionary aspect without dramatization, regarding it as an interesting psychical aberration of relative insignificance in a larger context:

> to a reader who can make due allowance in the report for the reporter's peculiarities, the results are still instructive, and a more striking testimony to the sublime laws he announced, than any that balanced dullness could afford.[258]

Emerson's principal objection is to Swedenborg's theological system, which, in his opinion, suppresses the symbolical interpretation of nature and eliminates all true individuality and freedom; we have already encountered similar reservations in Herder at the beginning of the romantic movement. Even what to Emerson was Swedenborg's greatest strength, namely, his gospel of love, suffers from this theological aridity. Though Emerson obviously read Swedenborg with considerable effort and with many reservations, the result was a tribute to the greatness of the author, generously and gracefully expressed in the conclusion of the essay: Swedenborg has given mankind an intimation of the innermost harmony and bliss of existence, and the circumstance that the radiance of what he had glimpsed had dazzled his own vision is only more powerful evidence of the glory of the drama.

Emerson's essay also opened the door to the world of Swedenborg for many people. For example, it was his fervent admiration for Emerson—the third most-quoted author in his oeuvre after Goethe and Nietzsche—that led the remarkable Swedish poet and aphorist Vilhelm Ekelund (1880–1949) to begin to read Swedenborg at the end of 1913.[259] For Ekelund's generation, however, there was a whole series of signposts leading to Swedenborg among the swarm of theosophical and spiritualistic works, which were essential to the change in spiritual climate at the end of the nineteenth century. In the English-speaking world, the poet who expressed this renaissance of mysticism most brilliantly was undoubtedly William Butler Yeats. However, his poetry has such profound and highly ramified roots in the hermetic tradition that it is almost impossible to isolate Swedenborg's influence; but Yeats himself declared in a letter in 1915 that his foremost mystical authorities were Boehme, Blake, and Swedenborg.[260]

In the general introduction to his writings as late as 1937 (published posthumously in 1961), Yeats avowed his belief in an "interpenetration of natural and supernatural" and mentioned Irish folklore, spiritism, Swedenborg, and Indian religions as sources of wisdom for future poets.[261] As William York Tindall and others assumed, it was probably Swedenborg's doctrine of correspondence that fascinated Yeats; but it is clear, not least from the many passing allusions, that Yeats was moved by the scenes from the world of spirits, by the visions of the eternally young and eternally loving angels.[262] Just after he received the Nobel Prize in Stockholm in 1923 and was looking at his diploma, which pictures a young man listening to a muse, Yeats immediately thought of Swedenborg; and he wrote in his autobiographical notes:

> I was good-looking once like that young man, but my unpracticed verse was full of infirmity, my Muse old as it were; and now I am old and rheumatic, and nothing to look at, but my Muse is young. I am even persuaded that she is like those Angels in Swedenborg's vision, and moves perpetually "towards the day-spring of her youth."[263]

It is unlikely that any of Swedenborg's countrymen in that august assemblage would have found the association as natural as Yeats did.

Yeats gained his insight into Swedenborg's theosophy in the course of his work on Blake's *Prophetic Books*, which he published in 1893; and, as Tindall pointed out, his confrontation with French symbolists did not change his thinking but confirmed conclusions he had already reached.[264] Other modern poets, however, became interested in Swedenborg through nineteenth-century French literature. When the young hero of Balzac's novel *Louis Lambert*, who has so many characteristics in common with the author, is introduced to Madame de Staël in 1811, he is avidly reading *Heaven and Hell.* Balzac stresses that in those days it was only Saint-Martin, the renowned illuminist, and a handful of other men of letters in France who had ever heard of Swedenborg.[265] This may be true of the period in question, but it does not mean that Swedenborg was unknown before that. On the contrary, his message had penetrated the French milieu as early as the 1770s through the intermediary of English and German writers, and many of his theosophical works were translated in whole or in part at an early stage.[266]

However, this early Swedenborgian influx coincided to a great extent with the syncretistic mysticism called Illuminism in France. The best known and most important of the mystical societies had its seat in Avignon, where a former librarian of the Prussian King Frederick II, Dom Pernety, among others, developed a form of theosophy in which Swedenborg's heavenly secrets were combined with elements of

Catholic mysticism, of hermetic tradition, and of alchemy. The result was a brew that, for a time, was extremely attractive to quite a few people during the turbulent years of the revolution, and as Robert Darnton has pointed out, Swedenborg played a significant part in the underground literature of the time.[267] A disciple of Swedenborg was even a member of the dreaded Committee of Public Safety, and the number of Swedenborg societies in France increased during the first years of the revolution.[268]

As in Stockholm, the associations with alchemists and practitioners of animal magnetism eventually had disastrous effects on Swedenborg's influence, and the first wave of French Swedenborgianism receded in the 1790s. Meanwhile, a new and more qualified interest in the Swedish theosophist developed around 1820, partly as a result of the first complete French translation of his spiritual treatises; and it is to this that we can trace the literary influence of Swedenborg on Victor Hugo, Lamartine, George Sand and—above all—Balzac.[269] The protagonist in *Louis Lambert* expresses the greatest admiration for Swedenborg's doctrines, which undoubtedly agreed with Balzac's own, at least in the 1830s. Swedenborg's theosophy came to represent a definitive synthesis of all the religions of history.[270] Thus, Balzac's declaration of allegiance to Swedenborgianism as a religion did not imply a break with the Catholic Church, but was an expression of his desire for synthesis, his longing for a God for whom the physical and the spiritual are one and the same.

In *Séraphita* (1835), Balzac moved one step further in the direction of Swedenborgianism than he had in *Louis Lambert.*[271] In the latter book, he propounded the concept of unity on a theoretical level, while in the former he gave it poetic form. The hermaphrodite Séraphitus-Séraphita, whose father Balzac identifies as a cousin of Swedenborg himself, is depicted against the background of a magnificent Norwegian fjord, together with three beings, each representing a different attitude: the skeptical priest Becker, his loving and devoted daughter Minna, and the disharmonious romantic Wilfrid. The protagonist symbolizes humanity in the stage of becoming angelic, and the theme of the novel is his/her transformation from mortal into angel. The aesthetic result of this attempt is questionable, and it is equally doubtful whether the interpretation of Swedenborg is based on a real understanding of his philosophy; but *Séraphita* is certainly the most remarkable manifestation of Balzac the mystic. The novel is important in evaluating its author's realism, since it illustrates his belief in the supreme unity, in the spiritual being represented in nature. With his strong physiological and psychological interests, Balzac is a true disciple of Swedenborg in the manner in which he experiences and expresses the conditions of humanity.[272]

It is also through Balzac that the strongest Swedenborgian impulses were channeled into French tradition and thence disseminated in many directions: we find traces of *Séraphita* everywhere in subsequent literature. It was probably through Balzac that Baudelaire began to read Swedenborg. This resulted in a temporary outburst of enthusiasm, and Swedenborg obviously played a major role in introducing the poet to mysticism. What primarily fascinated Baudelaire was the notion of correspondence, which became a cornerstone of his aesthetic metaphysics. As Michaud and others pointed out, even though Baudelaire did not actually create an aesthetic system, sensitive readers of his criticism and his poetry could not help discovering how close he came to mysticism in his literary concept of reality: the state of poetic creativity reveals the true reality, provides the contact with ideas and spiritual forces within and beyond the visible phenomena.[273]

Passages in one of his Poe essays (1857) reveal unequivocally that Swedenborg helped Baudelaire to orchestrate the theme:

> It is this admirable, this immortal instinct of the beautiful which makes us consider the earth and its spectacles as a revelation, as something in *correspondence* with Heaven. The insatiable thirst for everything that lies beyond, and that life reveals, is the most living proof of our immortality. It is at the same time by poetry and *through* music that the soul glimpses the splendors beyond the tomb; and when an exquisite poem brings us to the verge of tears, those tears are not the proof of excessive pleasure; they are rather evidence of an aroused melancholy, of a condition of nerves, of a nature which has been exiled amid the imperfect and which would like to

take possession immediately, on this very earth, of a revealed paradise.[274]

And it was also Baudelaire who gave the classical lyrical expression to the notion of correspondence in the sonnet *Correspondances*:

La Nature est un temple où de vivants piliers
Laissent parfois sortir de confuses paroles;
L'homme y passe à travers des forêts de symboles
Qui l'observent avec des regards familiers.

Comme de longs échos qui de loin se confondent
Dans une ténébreuse et profonde unité,
Vaste comme la nuit et comme la clarté,
Les parfums, les couleurs et les sons se répondent.

Il est des parfums frais comme des chairs d'enfants,
Doux comme les hautbois, verts comme les prairies,
—Et d'autres corrompus, riches et triomphants,

Ayant l'expansion des choses infinies,
Comme l'ambre, le musc, le benjoin et l'encens,
Qui chantent les transports de l'esprit et des sens.

Nature is a temple where living pillars
Let sometimes emerge confused words;
Man comes there over forests of symbols
Which watch him with intimate eyes.

Like those deep echoes that meet from afar
In a dark and profound harmony,
As vast as night and clarity,
So perfumes, colors, tones answer each other.

There are perfumes fresh as children's flesh
Soft as oboes, green as meadows,
And others, tainted, rich, triumphant,

Possessing the diffusion of infinite things,
Like amber, musk, incense and aromatic resin,
Chanting the ecstacies of spirit and senses.[275]

This exquisite tissue of symbols naturally contains many elements in addition to the idea of correspondence, primarily the idea of concordance between different sensory areas, or synesthesias, and its mood of sensual melancholy is far removed from Swedenborg's mathematical aspirations. It is not bizarre details that Baudelaire borrowed from the master, but Swedenborg's total vision as a poet; and he felt that his function as a disciple was to decode and translate nature's hieroglyphics—an expression he often uses, possibly to associate with Swedenborg's own terminology.

Baudelaire's way of utilizing Swedenborg in his poetic universe became, to a great extent, the norm for his successors in the symbolist generation. Swedenborg's name is frequently mentioned in the numerous manifestoes and declarations of principles—incidentally, the visionary is even encountered in Breton's first surrealistic manifesto in 1924—and it is usually the artistic application of the doctrine of correspondence that is called for. Most striking, however, is that Swedenborg is classified in the hermetical tradition in almost the same way as a century earlier.[276] The explanation is the renaissance of occultism, which occurred in the wake of hypnotic therapy and the psychology of suggestion and which was of central importance in French intellectual life around 1890. The frequent references to Swedenborg do not, however, reflect any profound study of his works, which, of course, differ radically from the efforts of the alchemists, the Rosicrucians, and the spiritualists.

It is, nevertheless, symptomatic that August Strindberg first became a follower of Swedenborg in the full tide of this Parisian occultism and under its direct influence. He himself relates in *Inferno* (1897) how he came to read Balzac's *Séraphita*, an overwhelming experience, even though he had already read it as a young man:

Indeed, it proved absolutely new to me, and now that my mind was prepared for it I was able to absorb the contents of this extraordinary

> book. I had never read any of Swedenborg's works (for in his own country—which is also mine—he is accounted a charlatan, a madman with a lubricious imagination), and I was seized with ecstatic admiration as I listened to the voice of this angelic giant of the previous century being interpreted to me by the most profound of all French geniuses.[277]

Strindberg is unquestionably the greatest of Swedenborg's readers among Swedish men of letters to appreciate him as an artist. There were others who were much better informed on Swedenborg's works than Strindberg, particularly writers belonging to the romantic movement, and at least one of them adhered periodically to Swedenborgianism, namely, Carl Jonas Love Almqvist (1793–1866). To him, Swedenborg's ideas of the lives of the dead, which he made use of in his own scenes from the spiritual world, were tragically decisive in his entire attitude toward humankind and toward life.[278]

However, none of the romantic disciples of Swedenborg in Swedish literature has won an international reputation comparable with Strindberg's. According to *Inferno*, it was through the intermediary of *Séraphita* that "Swedenborg revealed himself as a spiritual mentor in my life—in which he has played a tremendous role—and thus, on the anniversary of his death, he presented me with palms—whether of victory or martyrdom who could say?"[279] The date was March 29, 1896, but as usual Strindberg dramatized the incident: in reality, he probably first became acquainted with Swedenborg twenty years earlier, when he was working at the Royal Library in Stockholm, whose director Gustaf Klemming had published the *Journal of Dreams* and was the patron of the young library assistant August Strindberg.[280]

Reading Swedenborg's works, nevertheless, had a decisive effect on the development of the crisis that had plagued Strindberg for several years, a religious crisis that coincided with literary sterility and completely wild scientific speculations. Strindberg's suspiciousness developed during these turbulent years in the direction of clearly pathological delusions of persecution. His encounter with Swedenborg, primarily *Arcana Coelestia*, delivered him of the fear of madness, since he found a way of interpreting his experiences as consequences of the disciplinary activities of the spirits. He applied Swedenborg's term *vastatio*, devastation, to these years of imagined persecution, and he found his situation portrayed in detail in Swedenborg's *Journal of Dreams* and in the accounts of hell in *Arcana*. Swedenborg's post-mortem inferno was for Strindberg his own daily life in Paris and in the many other places to which he was driven during the miserable years of the mid-1890s.

Swedenborg's doctrine of activity, the doctrine that man must struggle for salvation with his evil nature, was a lifesaver for Strindberg, and he wrote in *Legends* (1898) that Swedenborg "has become my Virgil, who guides me through hell, and I follow him blindly."[281] But as could be expected, Strindberg also felt a need to free himself from this teacher, as he had done with all the other great spirits who had influenced his life and work. In the fragment entitled *Wrestling Jacob*, at the end of *Legends*, his book of confession, he attacks the man who had saved him from the madhouse, and he does so in order that his soul should not be injured by the elements in Swedenborg that are too human, too petty:

> So long as Swedenborg in the *Arcana* and the *Apocalypse* treats of revelations, prophecies, interpretations, he has a religious effect upon me, but when in the *Vera Religio* he begins to reason about dogmas, he becomes a freethinker and Protestant. When he draws the sword of reason, he has himself chosen the weapons, and they are likely to prove bad ones for himself. I wish to have religion as a quiet accompaniment to the monotonous music of life, but here it is a matter of professional religion and pulpit-discussion—in brief, a struggle for power.[282]

What terrified him, therefore, was mainly *Vera Christiana Religio*, and he bid a solemn farewell "with gratitude, as of one who, although with alarming pictures, had frightened me like a child back to God."[283]

This farewell, however, was not definitive. About ten years later, when Strindberg published the first volume of *A Blue Book*, he dedicated it to "Emanuel Swedenborg, mentor and master"; the superficial reason for this was that Swedenborg's remains were brought home to

Sweden and Uppsala Cathedral from London at the same time. Combined with the teachings of French occultism at the end of the nineteenth century, influences from Swedenborg play a very important part in this strange book of reflections; and according to contemporary notes in the so-called *Occult Diary*, Strindberg was still seeking relief from his own unhappiness in the theosophy of the visionary. It is clear that the complex, somber sense of life portrayed in Strindberg's later dramas with such extraordinary power had profound roots in the works of Swedenborg.

As a religious body, Emanuel Swedenborg's New Jerusalem has not achieved a large following. Nevertheless, his writings—and even more so his aspirations and the vision on which they are founded—are among the most remarkable in Western literature. The main purpose of my presentation has been to try to sketch the natural-philosophical background of Swedenborg's development and to indicate its roots in ancient and Christian tradition and its relative agreement with contemporary science. The portrait should have been given clearer contours, but it may nevertheless correspond to my high esteem for its subject, confirmed by a host of witnesses infinitely more competent than I. Through his ability to inspire writers who have exerted a major influence on modern literature—Balzac, Baudelaire, Yeats, Borges, Milosz, and in Swedenborg's ungrateful mother country, Strindberg, Vilhelm Ekelund, Gunnar Ekelöf, Lars Gyllensten, Kerstin Ekman, and Olof Lagercrantz—the scientist and mystic Emanuel Swedenborg is deserving of the place of honor accorded to him by Emerson among Representative Men.[284] And more than two centuries after Swedenborg at last was taken into the world of spirits that he described in such detail and in such a fascinating manner, his inspiration is still a stimulus to research, still provokes controversy, and still serves as a source of consolation.

Notes

Chapter One
"The Age and the Man," pages 3–20

1. *Oeuvres complètes de Voltaire*, vol. V (Paris, 1855), 674: "je dis que j'ai connu beaucoup de personnes que le cartésianisme a conduites à n'admettre d'autre Dieu que l'immensité des choses. . . ."; Aram Vertanian, *La Mettrie's L'homme machine. A Study in the Origin of an Idea* (Princeton, New Jersey: Princeton University Press, 1960), 49.

2. Rolf Lindborg, *Descartes i Uppsala. Striderna om "nya filosofien" 1663–1689* (Uppsala, Sweden: Almqvist & Wiksell, 1965); with an English summary "The Contentions about Cartesianism in Uppsala 1663–1689." Alfred H. Stroh interpreted the outcome of the Cartesian controversy as a victory for the philosophers; see "The Sources of Swedenborg's early Philosophy of Nature," in *Emanuel Swedenborg: Opera quaedam aut inedita aut obsoleta de rebus naturalibus* (abbr. *Opera*), published by the Royal Swedish Academy of Science, vol. III (Stockholm, 1911): xv–xxxvii.

3. *Oeuvres complètes de Baudelaire*, vol. II (Paris: Le club du meilleur livre, 1955), 478: ". . . le ciel est un très-grand homme*; que tout, forme, mouvement, nombre, couleur, parfum, dans le* spirituel *comme dans le* naturel, *est significatif, réciproque, converse,* correspondant."

4. Martin Lamm, *Swedenborg. En studie öfver hans utveckling till mystiker och andeskådare* (Stockholm: Hugo Geber, 1915), 23; in my opinion, this is still the most important work on Swedenborg, and fortunately, it is currently being translated into English; see the bibliography for information on other editions of this work. See also Ernst Benz, *Emanuel Swedenborg. Naturforscher und Seher* (Munich, Germany: Hermann Rinn, 1948), 30–52; Signe Toksvig, *Emanuel Swedenborg. Scientist and Mystic* (1948; reprint New York: Swedenborg Foundation, 1983), 43–55; Cyriel Odhner Sigstedt, *The Swedenborg Epic* (New York: Bookman Associates, 1952), 19–25; Inge Jonsson,

Swedenborgs korrespondenslära (Lund, Sweden: Almqvist & Wiksell, 1969), 29, 395; with an English summary, "Swedenborg's Doctrine of Correspondence."

5. Svante Lindqvist, *Technology on Trial: The Introduction of Steam Power Technology into Sweden, 1715–1736* (Uppsala, Sweden: Almqvist & Wiksell, 1984),158–170.

6. *Opera,* I (Stockholm, 1907), 321.

7. Ibid., 327. The Latin words might be rendered "extraordinarily acute."

8. *Resebeskrifningar af Emanuel Swedenborg 1710–1739,* published by the Royal Swedish Academy of Science, third edition (Stockholm, 1911). An English translation can be found in Rudolph L. Tafel, *Documents concerning the Life and Character of Emanuel Swedenborg,* vol. II, sec. pr.(London: The Swedenborg Society. 1890), 3–130.

9. *Swedenborg's Journal of Dreams 1743–1744,* trans. J.J.G. Wilkinson, commentary by Wilson Van Dusen (New York: Swedenborg Foundation, 1986). A new edition was recently published in Swedish, *Swedenborgs Drömbok,* edited by Lars Bergquist (Stockholm: Norstedts, 1988); an English translation of this version is scheduled for publication in 2000.

10. Tafel, *Documents* II, 215. The translation given there is somewhat inaccurate.

11. *Prodromus philosophiae ratiocinantis de infinito, et causa finali creationis: deque mechanismo operationis animae et corporis* (Dresden and Leipzig, 1734) (abbr. *De Infinito*), 268; English translation by J.J.G. Wilkinson, *The Infinite and the Final Cause of Creation* (London: The Swedenborg Society, 1908), 230.

12. Published in seven volumes by Immanuel Tafel, *Emanuelis Swedenborgii Diarium Spirituale* (Tübingen and London: W. Newbery, 1843–1846). English translation by George Bush, James Buss and John H. Smithson (London: James Speirs, 1883–1902). A second Latin edition was published in four volumes under the title *Experientiae spirituales* (Bryn Athyn, Pennsylvania: Academy of the New Church, 1983–1993).

Chapter Two
"Swedenborg's Philosophy of Nature," *pages* 21–33

13. *Opera,* I, 212: the Latin means "Newton's motion principles of the planets" and "planetary body."

14. *Mathematical Principles of Natural Philosophy*, Trans. Andrew Motte, *Great Books of the Western World*, vol. 34 (London: Encyclopaedia Britannica, Inc., 1952), 371.

15. *Opera,* III, 285, 303.

16. Ibid., 305.

17. *Opera,* I, 260.

18. Alfred H. Stroh, "Swedenborg's Early Life, Scientific Works, and Philosophy," *The New-Church Magazine* (1915): 400–403; and (1916): 13; Sigstedt, *Swedenborg Epic*, 116–117.

19. Tafel, *Documents*, II, 871.

20. *Opera,* III, 297–298 and 319–320.

21. "Notes on the Development and Texts of Swedenborg's Early Philosophy of Nature and 'Principia'," in *Opera* III, 324.

22. *Biographia Literaria*, chapter XIV; when Coleridge read *De Cultu et Amore Dei*, he made no comparisons with Burnet but referred with some hesitation to Buffon; see an unpublished Ph.D. dissertation by Leonard Martin Edmisten, *Coleridge's Commentary on Swedenborg*, University of Missouri, 1954, 235–239.

23. *De Cultu et Amore Dei*, n. 10; *The Worship and Love of God*, trans. Alfred H. Stroh and Frank Sewall (1925; rpt. West Chester, Pennsylvania; and London: Swedenborg Foundation and Swedenborg Society, 1996), 19–21. Swedenborg's own paragraph numbers will be used in citations wherever possible within the text; page numbers refer to translations.

24. *Principia rerum naturalium* (abbr. *Principia*), I:II:10; *The Principia*, trans. Augustus Clissold, vol. 1 (London: W. Newbery, 1846), 54.

Chapter Three
"From Inorganic to Organic Nature," pages 35–67

25. Jacques Roger has given a brilliant survey of this process, with the accent on France, in *Les sciences de la vie dans la pensée française du XVIIIe siècle* (Paris: Armand Colin, 1963), 161–453.

26. Joseph Needham, *A History of Embryology* (Cambridge University Press, 1934), 148.

27. *Daedalus Hyperboreus*, no. VI (1718), 13; facsimile edition in *Kungliga Vetenskaps Societetens i Upsala tvåhundraårsminne* (Memorial Publication

of the Royal Society of Science in Uppsala on its 200th Anniversary) (Uppsala: 1910).

28. In a letter to Benzelius in May 1720; *Opera* I, p. 304, translated by Alfred Acton in *Letters and Memorials of Emanuel Swedenborg* (abbr. *Letters*), vol. 1 (Bryn Athyn, Penna.: Swedenborg Scientific Association, 1948), 236–238.

29. *De Infinito,* 194; translated by Wilkinson, *Infinite . . . Cause,* 167–168.

30. Ibid., 247; Wilkinson translation, 212.

31. This essay has been translated by Alfred H. Stroh in *Scientific and Philosophical Treatises by Emanuel Swedenborg,* part II:1 (Bryn Athyn, Penna.: Swedenborg Scientific Association, 1905), 13–32.

32. Translated by Alfred Acton in *Psychologica, being Notes and Observations on Christian Wolff's Psychologia Empirica* (Philadelphia: Swedenborg Scientific Association, 1923), 78–79.

33. *Ms cod. 88–93,* pp. 496–497; translated by Alfred Acton in his preface to *Three Transactions on the Cerebrum, a posthumous work by Swedenborg,* vol. I (Philadelphia: Swedenborg Scientific Association, 1938), xxi.

34. Ibid., xxiii–xxiv.

35. See Carl Robsahm's *Memoirs* in Tafel, *Documents,* I, 35–36.

36. Paragraph 432; translated by J.J.G. Wilkinson, *Angelic Wisdom concerning the Divine Love and concerning the Divine Wisdom* (London: Swedenborg Society, 1885), 342–343.

37. Martin O. Ramström, "Swedenborg on the Cerebral Cortex as the Seat of Psychical Activity," in *Transactions of the International Swedenborg Congress 1910* (London: Swedenborg Society, 1912), 56–70. Torsten Gordh Jr. and Patrick Sourander, "Swedenborg, Linné och hjärnforskningen," in *Nordisk Medicinhistorisk Årsbok 1990,* 97–117 (with a short summary in English).

38. *The Brain considered Anatomically, Physiologically and Philosophically by Emanuel Swedenborg,* edited, translated, and annotated by R.L. Tafel, vol. I–II (London: James Speirs, 1882–1887).

39. Paragraph 1, *The Economy of the Animal Kingdom, considered Anatomically, Physically, and Philosophically by Emanuel Swedenborg,* translated by Augustus Clissold, vol. I (London: W. Newbery, 1845), 1. As observed by R.L. Calatrello, this is a too-literal rendering of the Latin title; it should be "The Organization of the Soul's Kingdom, i.e. the Body "(*animalis* = that which belongs to the *anima,* the soul); see Calatrello's doctoral thesis, "The Basic Philosophy of Emanuel Swedenborg, with Implications for Western Education," University of Southern California, 1966, p. 9 (microfilm).

40. See Acton's preface to *The Cerebrum*, I, xii-xxix, and Swedenborg's text, ibid., I, 21–31.

41 *Oeconomia,* II, paragraph 116; Clissold translation, II, 138.

42. *De Cultu et Amore Dei,* n. 95; Stroh and Sewall translation, *Worship and Love,* 224–225.

43. *Oeconomia,* II, paragraph 200; Clissold, II, 195.

44. Quoted from the introductory biographical note in the translation of *Principia mathematica* in *Great Books of the Western World* 34 (Encyclopaedia Britannica, 1952), x.

45. *Oeconomia* I, paragraph 253; Clissold, I, 230.

46. *De Cultu et Amore Dei,* n. 20, note m; Stroh and Sewall, 33–34.

47. Lamm, 69–70; Inge Jonsson, *Swedenborgs skapelsedrama De Cultu et Amore Dei* (with a summary in English: *Swedenborg's Drama of Creation De cultu et amore Dei*) (Stockholm: Natur och Kultur, 1961), 94–95, 309; Sturm, *Physica electiva sive hypothetica,* vol. III (Nuremberg, 1697), 149–157.

48. *Eléments de la philosophie de Newton, Oeuvres complètes,* vol. III, chapter I, 753.

49. Published in English translation by Alfred A. Acton, *A Philosopher's Note Book* (Philadelphia: Swedenborg Scientific Association, 1931); see bibliography.

50. See Acton's preface to *The Cerebrum*, xvii–xix; Toksvig, 88–95, and Sigstedt, 145. The arguments have been censured in my paper "Köpenhamn-Amsterdam-Paris: Swedenborgs resa 1736–1738" (with a summary in English: "Copenhagen-Amsterdam-Paris: Swedenborg's journey 1736–1738"), *Lychnos* (Annual of the Swedish History of Science Society) 1967–1968, 55–73, 73–76.

51. Roger, *Les sciences,* 446.

52. Ibid., 461–462.

53. *Tentamina Theodicaeae*; preface n. 29, Latin edition (Frankfurt and Leipzig, 1739), 414. In the French version, Leibniz uses the word *correspondance* but the Latin translation has *commercium.*

54. Needham, *Embryology,* 22, 61.

55. See, for example, Joseph Moreau, *L'Univers Leibnizien* (Paris: Emanuel Vitte, 1956), 171–190.

56. See Roger, 293–370.

57. Discussed by Roger, note 256, pp. 370–371.

58. *Histoire de Charles XII,* livre premier, *Oeuvres complètes,* vol. IV, p. 450.

59. Tore Frängsmyr, *Wolffianismens genombrott i Uppsala* (Uppsala: Almqvist & Wiksell, 1972), with a summary in English, "The Emergence of Wolffianism at Uppsala," 88–102.

60. *Principia,* III, appendix; Clissold, II, 363–367; Tafel, *Documents* II, 29. See also Hugo Lj. Odhner, "Christian Wolff and Swedenborg," *The New Philosophy* (1951), 237–251.

61. *Ms. cod. 57–104,* pp. 92–93; photolithographically published by R. L. Tafel in *Emanuelis Swedenborgii Autographa,* vol. VI (Stockholm, 1869), 318; printed in *Opera,* III,265–268, and in English translation by R. Brown and Alfred H. Stroh in *Scientific and Philosophical Treatises,* II:1, 57–60.

62. Brown and Stroh, 60.

63. Antony van Leeuwenhoek, *Arcana naturae detecta,* vol. I (Leyden, 1722), 10.

Chapter Four
"Ontology and Psychology," *pages* 69–91

64. *Oeconomia* I. n. 579; Clissold, II, 1–3.

65. Ibid., n. 584; Clissold, II, 5–7.

66. Christian Wolff, *Philosophia prima sive Ontologia, methodo scientifica pertractata* (abbr. *Ontologia*), (Frankfurt & Leipzig, 1730), §§ 746, 769, 770, 776; *Cosmologia generalis methodo scientifica pertractata* (abbr. *Cosmologia*), (Frankfurt & Leipzig, 1731), §§ 191, 192, 205, 213, 218, 378, 379; Scipion Dupleix, *Corps de philosophie, livre de la logique,* (Geneva, 1636), 43.

67. *Oeconomia* I, n. 649; Clissold, II, 50–51.

68. *Cosmologia,* §§ 182, 187, 227–246; *Oeconomia* I, nn. 629-630; Clissold, II, 31–33.

69. *Oeconomia* I, n. 630; Clissold, II,33.

70. Cf. *Oeconomia* I, n. 635; Clissold, II, 37; and Dupleix's *Corps de philosophie, livre de la physique,* 208—209, 216—217; *Oeconomia* I, n. 629; and Dupleix, 43-44 ; also *Oeconomia* II, n. 356, Clissold, II, 350; and Dupleix, 36–37.

71. In *Ms cod.* 54–113, pp. 235–256, Swedenborg made extensive excerpts on ontological matters from these sources; see vol. VI of the *Autographa* series, 323-342; English translation by Alfred Acton, *Ontology* (Boston: 1901).

72. *Nova acta eruditorum* (1735), 556–559.

73 *Oeconomia* II, n. 311; Clissold, II, 300.

74. *Oeconomia* I, chapter VIII, "An Introduction to Rational Psychology"; volume II, chapter III, "The Human Soul"; the fragmentary psychological transactions published by J.J.G. Wilkinson in *Emanuelis Swedenborgii Opuscula quaedam argumenti philosophici* (abbr. *Opuscula*), (London: W. Newbery, 1846), and translated by Alfred Acton in *Psychological Transactions by Emanuel Swedenborg* (Philadelphia: Swedenborg Scientific Association, 1920, reprt. 1955); the unfinished part VII of *Regnum animale* published by Immanuel Tafel, *De anima* (Tübingen & London: W. Newbery, 1849), translated by N.R. Rogers and Alfred Acton, *Rational Psychology* (Philadelphia: Swedenborg Scientific Association, 1950).

75. *De Anima*, 57–75; Rogers and Acton, *Rational Psychology*, 71–92.

76. Ibid., 255–256; Rogers and Acton, 318–319; the quotation is from Locke's *Essay concerning Human Understanding* volume IV, chapter XVII, 14.

77. *Oeconomia* I, nn. 1–28; Clissold, I, 1–15, particularly 9–10: "When, after a long course of reasoning, they make a discovery of the truth, straightway there is a certain cheering light, and joyful confirmatory brightness, that plays around the sphere of their mind; and a kind of mysterious radiation—I know not whence it proceeds, —that darts through some sacred temple in the brain. Thus a sort of rational instinct displays itself, and in a manner gives notice that the soul is called into a state of more inward communion, and has returned at that moment into the golden age of its intellectual perfections. The mind that has known this pleasure (for no desire attaches to the unknown), is carried away wholly in pursuit of it; and in the kindling flame of its love despises in comparison, as external pastimes, all merely corporeal pleasures. . . ."

78. Lamm, 64–68.

79. *Oeconomia,* II, nn. 275–277; Clissold, II, 253–260.

80. The famous translation by Pierre Coste, first edition 1700, second 1729.

81. *Oeconomia,* II, n. 282; Clissold, II, 265.

82. Lamm, 54–59; I will return to the problem of Neo-Platonism in Chapter 8.

83. *Regnum animale*, VI:2, published by Immanuel Tafel (Tübingen & London: W. Newbery, 1849), 23–46; translated by J.J.G. Wilkinson, *The Generative Organs* (London, 1852), 31–62.

84. *Oeconomia,* II, nn. 292–296; Clissold, II, 273–282.

85. *De Cultu et Amore Dei*, nn. 39–48; Stroh and Sewall, 63–82.

86. Ramström, 67–69.

87. *Oeconomia,* II, n. 297; Clissold, II, 282–284.

88. *Daedalus Hyperboreus* V; facsimile edition (Uppsala 1910) III,132.

89. *Oeconomia* II, n. 299; Clissold II, 285–286.

90. Moreau, 182–190, 237–244; cf. Jacques Jalabert, *Le Dieu de Leibniz* (Paris: Presses Universitaires de France, 1960), 191–201.

91. See note 53 to chapter 3 above.

92. *De Anima,* 75–81; Rogers and Acton, 92–99.

93. *Emanuelis Swedenborgii Oeconomia regni animalis in transactiones divisa* III, ed. by J.J.G. Wilkinson (London: W. Newbery, 1847); English translation by Alfred Acton (Philadelphia: Swedenborg Scientific Association, 1918).

94. Ibid., nn. 266a–268; Acton, 189–195.

95. Ibid., nn. 273–273a; Acton, 200–203. When Coleridge met with the doctrine of forms in *De Cultu et Amore Dei,* n. 6, he was very impressed: "it would of itself suffice to mark Swedenborg as a man of philosophic *genius,* radicative and evolvent" (Edmisten, 235).

96. *Opuscula,* 96–97; Acton, 27–28.

Chapter Five
"Universal Philosophy and Correspondence," pages 93–121

97. Published in *Opuscula,* 3–5; Acton., 7–10.

98. L. Couturat and L. Leau, *Histoire de la langue universelle* (Paris: Librairie Hachette, 1903), 11–22. For a detailed more recent survey, see Paolo Rossi, *Clavis universalis* (Milan, 1960), 179–236.

99. *De l'Esprit* I:IV, *Oeuvres complètes de C.-A. Helvétius,* vol. I (Paris, 1818), 38.

100. Moreau, 81–99; Rossi, 237–258.

101. See, for example, Bertrand Russell, *A Critical Exposition of the Philosophy of Leibniz,* second ed. (London: George Allen & Unwin, Ltd., 1949), 169–171.

102. Gustaf Eneström, "Emanuel Svedenborg såsom matematiker" (E.S. as Mathematician), Appendix to *Transactions of the Royal Swedish Academy of Science,* vol. XV, Stockholm, 1890.

103. *Three Transactions on the Cerebrum* I, nn. 603–605, pp. 337–341.

104. *Oeconomia* I, n. 650; Clissold II, 54–55.

105. Published in *Opuscula,* 91–122; Acton, 21–64.

106. *Principia* I:I, Clissold, I, 1–45.

107. *Oeconomia* I, n. 651; Clissold II, 55; Wolff's *Ontologia*, § 755: "Quoniam cognitio rerum mathematica consistit in cognitione quantitatis rerum; qualitates autem quantitatem habent, nempe gradus, ope figurarum & numerorum exacte atque distincte intelligibiles; *Qualitatum possibilem esse cognitionem mathematicam patet.*"

108. The quotation is to be found in Wolff's extensive commentary to the passage referred to in the note above. Coleridge unhesitatingly agreed to Swedenborg's demand for a mathematical philosophy of universals: "A most important suggestion, or rather an indispensable condition of all further progress in real science" (Edmisten, 217–218).

109. *Psychologia empirica methodo scientifica pertractata* (Frankfurt & Leipzig, 1732), §§ 294–312, pp.129–140.

110. The manuscript was published in vol. VI of the *Autographa* series, 265–269, and translated in *Scientific and Philosophical Treatises* II:1, 49–56.

111. *Oeconomia* II, n. 211; Clissold. II, 203–204. This passage was highly praised by Coleridge: "I remember nothing in Lord Bacon superior, few passages equal, either in depth of thought or in richness, dignity, and felicity of diction, or in weightiness of the truths contained, to the nn. 208 to 214 inclusive" (Edmisten, 224).

112. *De Anima*, 66; Rogers and Acton, *Rational Psychology*, 81.

113. See chapter three, "Works of the Years 1734–1745," page 53 above.

114. *De Anima*, 57–61; Rogers and Acton., 71–75.

115. Nicolas de Malebranche, *Recherche de la vérité* I:IV, notated in *ms. cod. 36-110*, see *A Philosopher's Note Book*, 305.

116. *Diarium Spirituale*, n. 2728; op.cit. II, 332. New edition by J.D. Odhner, *Experientiae Spirituales*, vol. II (Bryn Athyn, Penna.: Academy of the New Church, 1989), 527.

117. *De Anima*, 255–258; Rogers and Acton, 318–321.

118. Ibid., 63; Rogers and Acton., 78.

119. *Clavis Hieroglyphica arcanorum naturalium & spiritualium, per viam repraesentationum et correspondentiarum*, edited by Robert Hindmarsh, London, 1784; translated in *Psychological Transactions*, 157–213; see also the list of corrections of the Latin text, ibid., 281–282.

120. See pages 95–97 above.

121. *Clavis Hieroglyphica*, 9; *Psychological Transactions*, 166–167.

122. Ibid., 10; op.cit., 167–168.

123. *Regnum animale anatomice, physice et philosophice perlustratum*, vol. I (The Hague, 1744), n. 232, note u; English translation by J.J.G. Wilkinson, *The Animal Kingdom*, vol. I (London: W. Newbery, 1843), 451.

124. *Clavis Hieroglyphica*, 19–20; *Psychological Transactions*, 182–184.

125. *Arcana Coelestia*, n. 9011.

126. Erik Iversen, *The Myth of Egypt and its Hieroglyphs in European Tradition* (Copenhagen: Gad, 1961), 9–25.

127. See, for example, Mario Praz, *Studies in the Seventeenth Century Imagery* (London: Warburg Institute, 1939), 19–20, 155–185; and Albrecht Schöne, *Emblematik und Drama im Zeitalter des Barock* (München: C.H. Beck'sche Verlagsbuchhandlung, 1964), 17–59.

128. Francis Quarles, according to Iversen, 83.

129. Praz, 155.

130. *A Philosopher's Note Book*, 466.

131. *Arcana Coelestia*, n. 4966. In a very interesting essay on "Metaphysical Poetry and the Poetic of Correspondence" in his *Renaissance and Seventeenth Century Studies* (New York: Columbia University Press, 1964), 44–59, Joseph Anthony Mazzeo uses the term "the poetic of correspondence" to denote a view of metaphor held among the theorists of the conceit (Gracián, Tesauro, and others) and looks upon it as a precursor to the poetic of Baudelaire: "Although the treatises of the conceit appear to have been little read after the seventeenth century, the theory of metaphor which they developed was kept alive through the occult tradition, and reached Baudelaire through the agency of Swedenborg. It is not an accident that the great analogical complexity of much modern poetry should have been largely the work of Yeats and Baudelaire, two poets who studied the occult sciences and who revived the conception of the poet as one who approaches reality through the discovery of the analogies latent in nature" (Mazzeo, 58). It should be stressed, however, that Swedenborg never displayed any particular interest in poetics or the theory of metaphor; his doctrine of correspondence is primarily an offspring of another seventeenth-century tradition, the search for a universal language with the precision of mathematics.

132. *Psychologica*, 90–91.

133. Ibid., 92–94; cf. Wolff's *Psychologia Empirica*, 66–75.

134. *Clavis Hieroglyphica*, 24; op.cit., 192–193; *A Philosopher's Note Book*, 431.

135. Cf. Erich Auerbach, *Mimesis: The Representation of Reality in Western Literature*, translated by William Trask (New York: Doubleday Anchor Books, 1957), 13, 42–43, 64, 151–177.

136. Swedenborg's poetry in Latin and Swedish was published in *Emanuelis Swedenborgii Opera poetica* (Uppsala, 1910), cf., for example, 29–37, 57–62. All of Swedenborg's Latin poetry is now available in excellent critical editions by Hans Helander, *Festivus applausus in Caroli XII in Pomeraniam suam adventum* (Uppsala: Almqvist & Wiksell International, 1985), *Camena Borea* (Uppsala: Almqvist & Wiksell International, 1988); and *Ludus Heliconius and other Latin poems* (Uppsala: Almqvist &Wiksell International, 1995).

137. Cf. Wolff's *Theologia naturalis methodo scientifica pertractata*, part 1 (Frankfurt & Leipzig, 1736), § 539, pp. 486–487.

138. See *ms. cod. 79–115*, p. 37.

139. In a letter to the Landgrave of Hessen-Darmstadt in August, 1771: *Letters*, vol. II, 765.

140. *A Philosopher's Note Book*, 10.

141. Ibid., 334–339, 404, 407: quotations from *Recherche de la vérité* in a Latin translation, III:II:V, VI.

142. Ibid., 420–429, 431–433, 454–466, 477–480, 495–496, 499–500.

143. Hans Aarsleff, "Leibniz on Locke on Language," *American Philosophical Quarterly* 1 (1964): 2.

Chapter Six
"The Religious Crisis," *pages* 123–142

144. *Documents* II, 131–132.

145. Ibid., 148: the translation is corrected here.

146. Lamm, 132–133.

147. *Documents* II, 149.

148. Ibid., 154–161.

149. Ibid., 177–179.

150. Ibid., 181–185.

151. Ibid., 185.

152. Ibid., 187–188.

153. Lars Bergquist, *Swedenborgs Drömbok. Glädjen och det stora kvalet*, (Stockholm: Norstedts, 1988), 29–35; *Documents* II,. 160–161, 196, 201,

213–214; *Diarium Spirituale*, nn. 4749 m, 4767 m, 4774 m, 4785, 4791–4810, 4813–4817, 5886, 5988–5990, 5993, 5995, 6017, 6043, 6081, 6102; op.cit. IV, pp. 66-68, 79–80, 82–83, 202–203, 205–211, 212–214; V,. 86, 138–140, 141–144, 156–157, 180–181, 201–202, 219–220.

154. *Documents* II, 204, 208–209, 215.

155. Ibid., 196–197, 207–208.

156. *Regnum Animale*, vol. III (London, 1745), nn. 470–486; op.cit. II, 1844, pp. 463–523.

157. *Documents* II, 209.

158. Ibid., 209–215.

159. Ibid., 217–219.

160. In a letter to Gabriel Beyer in November 1769, *Documents* II, 280–282.

161. A survey of the arguments is given in my paper in *Lychnos* 1967–68: 48–51, 74–75.

162. Emil A.G. Kleen, *Swedenborg. En lefnadsskildring*, vol. II (Stockholm: Sandberg, 1920), 699–763; cf. Toksvig, 156–167.

163. Lamm, 196–225 (p. 201: "In Swedenborg's time, people saw only two possibilities to explain such religious views as they did not like, fraud or madness.").

164. See particularly the dreadful descriptions of Erik Benzelius, *Diarium Spirituale*, nn. 4851, 5148, 5722, 5751, 6016; Bush, et al, IV, 235, 322–323; V, 22, 32–33, 156.

165. *Documents* II,. 613–692; Tafel's account of the sources is valuable, but I cannot see that his conclusions stand scrutiny. An examination of Kant's opinions of Swedenborg can be found in a well-informed doctoral thesis by Robert Kirven, *Emanuel Swedenborg and the Revolt against Deism* (microfilm Brandeis University, 1965), 46–65.

166. N. 53, note p; Stroh and Sewall, 96–97.

167. N. 95; cf. chapter three.

168. N. 112; Stroh and Sewall, 250.

169. N. 122; ibid., 265.

170. N. 78; ibid., 188–190.

171. See my book on *De Cultu et Amore Dei*, 1961, 304–318.

172. *Opera* I, 337; *Letters* I, 499–500.

173. See Carl Robsahm's memoirs, *Documents* I, 34–36.

174. *Eman. Swedenborgii Adversaria in Libros Veteris Testamenti*, ed. by Immanuel Tafel, vol. I (Tübingen & London: W. Newbery, 1847), 7–8;

translated by Alfred Acton in *The Word of the Old Testament Explained*, vol. I (Bryn Athyn, Penna.: Academy of the New Church, 1928), 10–11.

175. Ibid., 25; Acton translation, 30, 33: "These things are premised. . . . But let us examine the Scriptures, especially with the purpose of searching the kingdom of God; that is to say, its future quality, and many things appertaining to it. The Scriptures treat of the kingdom of God, not here and there, but everywhere; for this kingdom was the end in the creation of all things both of heaven and of earth."

176. See Acton's most valuable *Introduction to The Word Explained* (Bryn Athyn, Penna.: Academy of the New Church, 1927), 124–130.

Chapter Seven
"The Secrets of Heaven," pages 143–173

177. *Documents* II, 248–249, 252–260, 268–271; see also Ernst Benz, *Swedenborg in Deutschland* (Frankfurt am Main: Vittorio Klostermann, 1947), 3–232, and Kirven, 65–90. See also Michael Heinrichs, *Emanuel Swedenborg in Deutschland. Eine kritische Darstellung der Rezeption des schwedischen Visionärs im 18. und 19. Jahrhundert* (Frankfurt: Europäische Hochschulschriften Reihe XX, vol. 47, 1979), passim.

178. Lamm, 209–225; Henri de Geymüller, *Swedenborg und die übersinnliche Welt*, translated from French by Paul Sakmann (Stuttgart: Deutsche Verlags-Anstalt, 1936), 310–316; Ernst Arbman, *Ecstacy or Religious Trance*, part I (Stockholm: Svenska Bokförlaget, 1963), 113; Toksvig, 156–229; Eric Dingwall, *Very Peculiar People* (New York: University Books, 1962), 11–68; *Adversaria in Libros Veteris Testamenti*, nn. 943, 1351–1353; Acton translation II, 245–246, 409–410.

179. *De Coelo et Ejus Mirabilibus et de Inferno* (abbr. *De Coelo*) (London, 1758), n. 248; English translation, *Heaven and its Wonders and Hell* (Harmondsworth: Penguin Books, 1938), 110–111; and *Heaven and Hell*, trans. George F. Dole (New York: Swedenborg Foundation, 1976).

180. *Adversaria in Libros Veteris Testamenti* I, nn. 14–16; Acton I, 19–20.

181. *Diarium Spirituale*, nn. 680–681, 2896–2898; Bush, et al, translation I, 232–233; II, 383–384.

182. Ibid., n. 5595; Bush et al IV, 470.

183. Ibid., nn. 5578–5597; Bush et al IV, 462–471.

184. Caspar Neumann, *Clavis Domus Heber, referans Januam ad Significationem Hieroglyphicam Literaturae Hebraicae perspiciendam* (Breslau, 1712), 145–146.

185. *Arcana Coelestia*, n. 793; English translation, *The Heavenly Arcana* (London: Swedenborg Society, 1861), vol. I, 284–285; also *Arcana Coelestia*, 12 vols., trans. John Clowes, ed. John F. Potts, 2nd edition (West Chester, Penna.: 1995–1998). *De Coelo*, n. 241; 1938 translation, 107–108.

186. N. 254; Bush et al, I, 68–69.

187. Johann Gottfried von Herder, *Adrastea und das achtzehnte Jahrhundert, Sämmtliche Werke*, ed. J. von Müller, vol. XXX (Stuttgart & Tübingen, 1829),122.

188. *Diarium Spirituale*, n. 866; cf. n. 4671; Bush et al I, 84; IV, 151; *De Amore Conjugiali*, n. 115; English translation by W. F. Wunsch, *Conjugial Love*, 2nd edition (West Chester, Penna.: Swedenborg Foundation, 1998).

189. *Diarium Spirituale*, nn.4865–4871; Bush et al IV, 239–241.

190. *Arcana Coelestia*, nn. 168–189, continued nn. 314–319; cf. *Diarium Spirituale*, nn. 1092–1109; Bush et al I, 348–352. See, for example, Elisabeth Kübler-Ross, *The Wheel of Life. A Memoir of Living and Dying* (New York: Station Hill Press, 1997), passim.

191. *Arcana Coelestia*, nn. 320–323, 443–459, 537–553, 684–691.

192. Ibid., nn. 692–700, 814–831, 938–969.

193. Ibid., nn. 1273–1278, 1376–1382.

194. Ibid., nn. 6695–6702, 6807–6817, 6921–6932, 7069–7079, 7170–7177 (Mercury); 7246–7254 (Venus); 7358–7365, 7475–7487, 7620–7622, 7742–7751 (Mars); 7799–7813, 8021–8032, 8111–8119, 8242–8251, 8371–8386, 8541–8547, 8627–8634, 8733–8741, 8846–8852 (Jupiter); 8947–8957, 9104–9111 (Saturn); 9232–9238 (the moon); 9438–9442, 9578–9584, 9693–9700, 9790–9795, 9967–9973, 10159–10166, 10311–10317, 10377–10385, 10513–10518, 10585–10590, 10708–10713, 10734–10739, 10751–10759, 10768–10772, 10783–10788, 10808–10814, 10833–10837 (the starry heaven).

195. Ibid., nn. 3624–3649, 3741–3750, 3883–3896, 4039–4055, 4218–4228, 4318–4331, 4403–4421, 4523–4534, 4632–4634, 4652–4660, 4791–4806, 4931–4953, 5050–5062, 5171–5190, 5377–5396, 5552–5573, 5711–5727.

196. Ibid., nn. 9350–9362.

197. *De telluribus in mundo nostri solari, quae vocantur planetae* (London, 1758); English translation, *The Earths in Our Solar System* (London:

Swedenborg Society, 1909); also *Earths in the Universe* in Emanuel Swedenborg, *Miscellaneous Theological Works,* trans. John Whitehead (West Chester, Penna.: Swedenborg Foundation, 1996), 425–551. A detailed comparison between his use of the materials in *Diarium Spirituale* of the Mercury spirits in *Arcana Coelestia* and this leaflet can be found in my book on Swedenborg's doctrine of correspondence (1969), 374–375.

198. *Diarium Spirituale,* n. 765; Bush et al I, 258–259.

199. *Arcana Coelestia,* nn. 2117–2133

200. *Vera Christiana Religio,* nn. 753–775; English translation, *The True Christian Religion* (London: Everyman's Library, 1933), 791–807; and *True Christian Religion,* 2 vol., trans. John C. Ager, 2nd edition (West Chester, Penna.: 1997).

201. Ibid., n. 791.

202. A complete list of the memorabilia in *Apocalypsis Revelata, De Amore Conjugiali* and *Vera Christiana Religio* is presented in my book on Swedenborg's doctrine of correspondence, 381–390.

203. *De Amore Conjugiali,* nn. 2–25.

204. Ibid., n. 41.

205. Sigstedt, 243, 463–464.

206. *Arcana Coelestia,* n. 553.

207. *De Amore Conjugiali,* nn. 45–54.

208. Ibid., n. 75.

209. Ibid., n. 281–286.

210. The arguments have been summarized in *De Coelo,* nn. 170–176.

Chapter Eight
"The Lord and His Word," *pages* 175–193

211. Harry Lenhammar, *Tolerans och bekännelsetvång* (with a summary in English: *Toleration and Doctrinal Unity: A Study in Swedish Swedenborgianism 1765–1795*), (Uppsala: Acta Universitatis Upsaliensis, 1966), 16.

212. *Letters* II, 630; *Documents* II, 260–262. Swedenborg's first known reader in England, Stephen Penny of Dartmouth, believed that the anonymous author of *Arcana Coelestia* might be William Law; see Kirven, 114

213. *Diarium Spirituale,* nn. 3485–3487, 3890, 3891, 5962; Bush et al III, 80–81, 222–223; V, 126–127. The problem is discussed at some length in my book on Swedenborg's doctrine of correspondence, 256–258.

214. *Adversaria in Libros Veteris Testamenti* I, n. 505: Acton translation I, 430–431.

215. Acton, *Introduction to the Word Explained*, 120–130.

216. Henri de Lubac, *Exégèse médiévale, Les quatre sens de l'Ecriture*, vol.. I–II (Paris: Aubier, 1959–1964).

217. *Diarium Spirituale*, nn. 1602–1607; Bush et al II, 15–16.

218. F.W. Farrar, *History of Interpretation* (London, 1885), 128–129: Swedenborg's reading of *De Doctrina Christiana* resulted in some excerpts, *A Philosopher's Note Book*, 101, 148.

219. de Lubac, I, 198–199.

220. Ibid., I, 23: the meaning of the letter teaches us what has occurred, the allegory what we should believe, the moral what we should do, the anagogical where we shall go. The origin of the rhyme dates, in fact, back to Augustinus de Dacia (about 1260), according to de Lubac.

221. *Historia Creationis a Mose Tradita*, in *Adversaria in Libros Veteris Testamenti* I, 1–25; Acton I, 3–30.

222. Ibid., n. 15; Acton I, 47.

223. *Doctrina Novae Hierosolymae de Scriptura Sacra* (Amsterdam, 1763); English translation *The Doctrine of the New Jerusalem concerning the Sacred Scripture* (London: Swedenborg Society, 1909); also in *Posthumous Theological Works,* vol. 2, trans. John Whitehead, 2nd edition (West Chester, Penna.: Swedenborg Foundation, 1996), 235–347.

224. Ibid., n. 21.See also *Vera Christiana Religio*, n. 202.

225. Ibid., n. 24; ibid., n. 206.

226. Ibid., nn. 85–87. See also *Arcana Coelestia*, n. 4105.

227. Ibid., nn. 98–117; *Vera Christiana Religio*, nn. 261–276.

228. *Apocalypsis Revelata* (Amsterdam, 1766); English translation by T.B. Hayward, *The Apocalypse Revealed* (Philadelphia, 1912), 49–50; also *Apocalypse Revealed,* 2 vols., trans. John Whitehead, 2nd edition (West Chester, Penna.: 1997). See also *Diarium Spirituale*, n. 6077; Bush et al V, 199–200.

229. *De Amore Conjugiali*, n. 77; *Vera Christiana Religio*, nn. 266, 279.

230. See, for example, the great encyclopaedia, *Encyclopédie, ou Dictionnaire raisonné des sciences, des arts et des métiers*, vol. 32 (Bern and Lausanne, 1780), 772.

231. Paul Hazard, *La crise de la conscience européenne* (Paris: Boivin & Cie, 1935), 3–29; see also J. Ho's Zürich dissertation *Quellenuntersuchung zur Chinakenntnis bei Leibniz und Wolff* (Hong Kong, 1962), passim.

232. Anders Hallengren, "Tartariets hemlighet" in *Öarna under vinden* (Stockholm: Bokförlaget Åsak, 1997), 26–47; English translation "The Secret of Magna Tartaria," *Gallery of Mirrors: Reflections of Swedenborgian Thought* (West Chester, Penna.: Swedenborg Foundation, 1998), 17–41.

233. *Vera Christiana Religio*, nn. 18–24. See also Lamm, 241–295.

234. Ibid., nn. 36–47.

235. Ibid., nn. 81–109.

236. Ibid., nn. 49–70.

237. *De Amore Conjugiali*, n. 444. See also Lamm, 261–262.

238. *Vera Christiana Religio*, nn. 114–133, 466–502.

239. Ibid., n. 508.

Chapter Nine
"New Jerusalem in the World," pages 195–210

240 *Documents* II, pp. 402-405; quoted at length by Sigstedt, 312–315.

241. *Svenska Mercurius* (Stockholm, 1764), 651.

242. *Documents* II, 395–397.

243. Lenhammar, 20–29.

244. *Documents* II, 405–416.

245. Lenhammar, 40–172, 324–330.

246. See Karin Johannisson, *Magnetisörernas tid: Den animala magnetismen i Sverige* (diss. Uppsala, 1974), Jan Häll, *I Swedenborgs labyrint. Studier i de gustavianska swedenborgarnas liv och tänkande* (diss. Uppsala 1995) and Jakob Christensson, *Lyckoriket. Studier i svensk upplysning* (diss. Lund 1996), all with summaries in English.

247. Lenhammar,175–321, 330–334.

248. Benz, *Swedenborg in Deutschland*, 235–285; see also Kirven, 46–65.

249. Benz, "Swedenborg und Lavater", *Zeitschrift für Kirchengeschichte*, 1938, 156–157; *Documents* II, 264–266, 277–278; Kirven, 90–95.

250. See, for example, Friedemann Horn, *Schelling and Swedenborg: Mysticism and German Idealism*, trans. George F. Dole (West Chester, Penna.: Swedenborg Foundation, 1997), Michael Heinrichs, *op.cit.*, and Inge Jonsson, "Swedenborg i Tyskland," in *Bland böcker och människor* (Uddevalla, 1983). Kirven's examination of Swedenborg's influences in Germany until 1840 is a very valuable contribution, 259–295.

251. Julius Langbehn, *Rembrandt als Erzieher*, 67–71 ed. (Leipzig: C.L. Hirschfeld, 1922),130; see also 264–289.

252. Walser alludes to Swedenborg's recounting of his vision in London in 1745, when he was admonished not to eat so much, and he gives it a cruel, ironic touch in a conversation in which a young woman become interested in the story only as a curious example of dietics (Swedish translation, 1981, 32–33).

253. *Documents* II, 797–834.

254. George Trobridge, *Swedenborg. Life and Teaching*, 4th rev. ed. (New York: Swedenborg Foundation, 1955), 210–226; and revised by R. H. Tafel Sr. and R. H. Tafel Jr, 1992, 139–150. See also Freda G. Griffith, *The Swedenborg Society 1810–1960* (London: Swedenborg Society, 1960); S.C. Eby, *The Story of the Swedenborg Manuscripts* (New York: New-Church Press, 1926).

255. Marguerite Beck Block, *The New Church in the New World: A Study of Swedenborgianism in America* (rpt. New York: Swedenborg Publishing Association, 1984), 73–74.

256. H.M. Margoliouth, *William Blake* (Oxford University Press, 1951), 74–78. See also a penetrating analysis by David V. Erdman, "Blake's Early Swedenborgianism: A Twentieth-Century Legend," *Comparative Literature* 5 (1953): 247–257. See also Harvey F. Bellin and Darrell Ruhl, ed., *Blake and Swedenborg: Opposition is True Friendship* (New York: Swedenborg Foundation, 1985) and my review in *Blake: An Illustrated Quarterly* 21, no 4: 158–160.

257. Leon Edel, *Henry James*, vol. I (London: Rupert Hart-Davis, 1953), 34–36, 38, 47; vol. II (1962), 141–142. See also Paul J. Croce, "A Scientific Spiritualism: The Elder James's Adaptation of Emanuel Swedenborg" in *Swedenborg and His Influence* (Bryn Athyn, Penna.: The Academy of the New Church, 1988), 251–262; and Dwight W. Hoover, "The Influence of Swedenborg on the Religious Ideas of Henry James, Senior," ibid., 263–276.

258. Ralph Waldo Emerson, *Representative Men* (London, 1901), 91.

259. Sven Lindqvist, *Dagbok och diktverk. En studie i Vilhelm Ekelunds Nordiskt och klassiskt* (diss. Stockholm: Bonniers, 1966) (with a summary in German), 236. I have generously been given the frequency figures from an ongoing computer concordance project in Lund, partly financed by the Royal Academy of Letters, History, and Antiquities.

260. Allan Wade, ed., *The Letters of W.B. Yeats* (London: Rupert Hart-Davis, 1954), 592.

261. David R. Clark, "Metaphors for Poetry: W.B. Yeats and the Occult" in *The World of W.B. Yeats*, ed. by Robin Skelton and Ann Saddlemeyer (Victoria, British Columbia, Canada: University of Victoria Press,1965), 62–63.

262. William York Tindall, "The Symbolism of W.B. Yeats" in *Yeats. A Collection of Critical Essays*, ed. by John Unterecker (Englewood Cliffs, New Jersey: Prentice-Hall, 1963), 46.

263. W. B. Yeats, *Autobiographies* (London: Macmillan & Co., Ltd., 1955), 541.

264. Tindall, 43–51.

265. *Oeuvres complètes de H. de Balzac*, vol. XVII (Paris: Michel Lévy Frères, 1870), 7.

266. Auguste Viatte, *Les Sources occultes du romantisme*, vol. I (Paris: Librairie Ancienne Honoré Champion, 1928), 80–103. See also Kirven, 217–253.

267. Robert Darnton, *Mesmerism and the End of the Enlightenment in France* (Cambridge: Harvard University Press, 1968), and *The Literary Underground of the Old Regime* (ibid., 1982): see Lynn R. Wilkinson, *The Dream of an Absolute Language: Emanuel Swedenborg and French Literary Culture* (Albany: State University of New York Press, 1996), 52–53.

268. Karl-Erik Sjödén, *Swedenborg en France* (Stockholm: Almqvist & Wiksell International, 1985), 23–35.

269. Viatte vol. II, 248–253; Sjödén, 154–178.

270. Balzac, *Oeuvres complètes*, vol. XVII, 68.

271. Ibid., 103–236.

272. See René Wellek, *Concepts of Criticism* (New Haven: Yale University Press, 1964),174–175. See also Samuel Rogers, *Balzac and the Novel* (Madison: The University of Wisconsin Press, 1953), 82–91; and Wilkinson, 147–216.

273. Guy Michaud, *Message poétique du symbolisme* (Paris: Librairie Nizet, 1951), 67–78; Wilkinson, 217–247.

274. Baudelaire, vol. I, 628–648; the quoted passage, p. 645, recurs in his Gautier essay 1859, vol. II, 94–95. Quoted here from *Baudelaire on Poe*, translated and edited by Lois and Francis E. Hyslop, Jr. (State College, Penna.: Bald Eagle Press, 1952), 140–141.

275. Baudelaire, vol. I, 674. Here quoted from *Baudelaire Selected Poems* with translations by G.Wagner (London: The Falcon Press, 1946), 27.

276. Guy Michaud, *La doctrine symboliste* (Paris: Librairie Nizet, 1947), 20–22, 40–41, 43.

277. *Inferno, Alone and Other Writings by August Strindberg in New Translations*, ed. Evert Sprinchorn, trans. Derek Coltman and Evert Sprinchorn (New York: Anchor Books, 1968),164.

278. Cf. Staffan Björck, "C.J.L. Almqvist: Romantic Radical" in *The American-Scandinavian Review* LVII, no 1 (1969): 24: "During the second decade of the 1800's he belonged to several societies of this type, in which Swedenborgian thought was united with a worship of the Old Norse–related to the Gothic vogue. . . ."; see also Olle Hjern, "Carl Jonas Almqvist: Great Poet and Swedenborgian Heretic," in *Swedenborg and His Influence*, 79–90.

279. Strindberg, 164.

280. Gunnar Brandell, *Strindbergs Infernokris* (diss Stockholm: Bonniers, 1950), 263; see also Göran Stockenström, "'The Great Chaos and the Infinite Order': The Spiritual Journeys of Swedenborg and Strindberg," in *Swedenborg and His Influence*, 47–78.

281. *Legends: Autobiographical Sketches by August Strindberg* (London, 1912), 108.

282. Ibid., 231.

283. Ibid., 233.

284. One of the most interesting contributions to Swedenborg literature in later years is Olof Lagercrantz' *Dikten om livet på den andra sidan. En bok om Emanuel Swedenborg* (Stockholm: Wahlström & Widstrand, 1996). The author, who is one of Sweden's most prominent critics, presents Swedenborg's theosophy as a colossal poem on a more dignified human life. Among other signs of an increasing Swedenborg vogue in Sweden may be mentioned that Swedenborg's *De Cultu et Amore Dei* was staged in September 1998 as an oratorium by a theatre company in the provinces led by the young actress Maria af Malmborg.

Selected Bibliography

Primary Sources

The following chronological list includes only Swedenborg's main works. For details the reader is referred to James Hyde, *A Bibliography of the Works of Emanuel Swedenborg* (London: Swedenborg Society, 1906), and to Greta Ekelöf and Alfred H. Stroh, *Kronologisk förteckning öfver Swedenborgs skrifter* (Chronological List of the Writings of Emanuel Swedenborg) (Uppsala: Royal Swedish Academy of Science, 1910). Also helpful is William Ross Woofenden, *Swedenborg Researcher's Manual* (Bryn Athyn, Penna.: Swedenborg Scientific Association, 1988), a work that is currently being revised. Although most of Swedenborg's works are available in many different translations and in a number of editions, for his theological works, I have listed the latest editions from the Swedenborg Foundation.

Opera poetica. Uppsala, 1910. Swedenborg's poetry has recently been published by Hans Helander in three excellent critical editions, *Festivus Applausus in Caroli XII in Pomeraniam suam adventum* (Uppsala, 1985), *Camena Borea* (1988), and *Ludus Heliconius and other Latin poems* (1995).

Daedalus Hyperboreus (The Northern Inventor). Uppsala, 1716–1718 (only partially translated into English).

Regelkonsten författad i tio böcker (The Science of Algebra).Uppsala, 1718.

Om Jordenes och Planeternas Gång och Stånd. Skara, 1718. New edition in *Emanuelis Swedenborgii Opera quaedam aut inedita aut obsoleta de rebus naturalibus,* vol. III. Stockholm, 1911. English translation, *Motion and Position of the Earths and Planets.* London: James Speirs, 1900.

Prodromus principiorum rerum naturalium. Amsterdam, 1721. New edition in *Opera* III (cf. above). English translation, *Chemistry.* London, 1847.

Miscellanea observata circa res naturales. Leipzig, 1722. New edition in *Opera* I (1907). English translation, *Miscellaneous Observations.* London, 1847.

Principia rerum naturalium sive novorum tentaminum phaenomena mundi elementaris philosophice explicandi. Dresden and Leipzig, 1734. New edition (phototyp.) Basel: Swedenborg Institut, 1954. English translation, *The Principia* I–II. London: W. Newbery, 1846. Later reprints.

Prodromus philosophiae ratiocinantis de infinito, et causa finali creationis: deque mechanismo operationis animae et corporis. Dresden and Leipzig, 1734. English translation, *The Infinite and the Final Cause of Creation.* London: Swedenborg Society, 1908.

Three Transactions on the Cerebrum. A Posthumous Work by Emanuel Swedenborg. Translated and edited by Alfred Acton. Vols. I–II. Philadelphia: Swedenborg Scientific Association, 1938, 1940.

Oeconomia Regni Animalis in transactiones divisa. Vols. I–II. London and Amsterdam, 1740–1741. English translation, *The Economy of the Animal Kingdom.* Vols. I–II. London: W. Newbery, 1845, 1846. Available in modern reprint.

Oeconomia Regni Animalis in transactiones divisa. Vol. III. Edited by J. J. G. Wilkinson. London: W. Newbery, 1847. English translation, *The Fibre.* Philadelphia: Swedenborg Scientific Association, 1918.

The Brain considered anatomically, physiologically and philosophically by Emanuel Swedenborg. Translated and edited by R.L. Tafel. Vols. I–II. London: James Speirs, 1882, 1887.

Regnum Animale anatomice, physice et philosophice perlustratum. Vol. I–II. The Hague, 1744; vol.. III. London, 1745. English translation, *The Animal Kingdom.* Vols. I–II. London: W. Newbery, 1843, 1844. Available in modern reprint.

Regnum Animale. Vol. IV. Edited by Im. Tafel. Tübingen and London, 1848. English translation, *The Five Senses.* Philadelphia: Swedenborg Scientific Association, 1914.

Regnum Animale. Vol. VI:2. Edited by Im. Tafel. Tübingen and London, 1849. English translation, *Generation.* London: W. Newbery, 1852. Available in modern reprint.

Regnum Animale. Vol. VII, *De anima.* Edited by Im. Tafel. Tübingen and London, 1849. English translation, *Rational Psychology.* Philadelphia: Swedenborg Scientific Association, 1950.

Selected Bibliography

Clavis Hieroglyphica arcanorum naturalium & spiritualium, per viam repraesentationum et correspondentiarum. Edited by R. Hindmarsh. London, 1784. English translation in *Psychological Transactions by Emanuel Swedenborg.* Philadelphia: Swedenborg Scientific Association, 1955.

De Cultu et Amore Dei. London, 1745. English translation, *The Worship and Love of God.* Trans. Alfred H. Stroh and Frank Sewall. 1914; rpt. West Chester, Penna., and London: Swedenborg Foundation and Swedenborg Society, 1996.

Adversaria in libros Veteris Testamenti. Vols. I–VI. Edited by Im. Tafel. Tübingen and London, 1847–1854. English translation, *The Word of the Old Testament Explained.* Vols. I–VIII. Bryn Athyn: Academy of the New Church, 1928–1948.

Diarium Spirituale. Vols. I–VII. Edited by Im. Tafel. Tübingen and London, 1843–1846. English translation, *The Spiritual Diary of Emanuel Swedenborg.* Vols. I–V. London: James Speirs, 1883–1902. Second Latin edition, *Experientiae Spirituales.* Vol. I–IV. Bryn Athyn: Academy of the New Church, 1983–1993. Two more volumes in preparation.

Arcana Coelestia quae in Scriptura Sacra seu Verbo Domini sunt detecta. Vol. I–VIII. London, 1749–1756. Two later editions. English *Arcana Coelestia.* Translated by John Clowes. Edited by John F. Potts. 12 vols. Second edition. West Chester, Penna.: Swedenborg Foundation: 1995–1998.

De Coelo et ejus mirabilibus, et de inferno. London, 1758. Several English translations, including (1) *Heaven and its Wonders and Hell.* London: Penguin Books, 1938; (2) *Heaven and Hell.* Translated by John C. Ager. Second edition. West Chester, Penna.: Swedenborg Foundation, 1995; and (3) *Heaven and Hell.* Translated by George F. Dole. West Chester, Penna.: Swedenborg Foundation, 1976.

De Nova Hierosolyma et ejus doctrina coelesti. London, 1758. English translation, *The New Jerusalem and its Heavenly Doctrine.* In *Miscellaneous Theological Works.* Translated by John Whitehead. Second Edition. West Chester, Penna.: Swedenborg Foundation, 1996. A recent translation is *The Heavenly City: A Spiritual Guidebook.* Translated by Lee Woofenden. West Chester, Penna.: Swedenborg Foundation, 1993.

De Telluribus in mundo nostri solari, quae vocantur planetae. London, 1758. English translation, *The Earths in the Universe.* In *Miscellaneous Theological Works.* Translated by John Whitehead. Second edition. West Chester, Penna.: Swedenborg Foundation, 1996.

De Ultimo Judicio, et de Babylonia destructa. London, 1758; *Continuatio de ultimo judicio.* Amsterdam, 1763. English translation, (1) *Concerning the Last Judgment; A Continuation of the Last Judgment.* London: Swedenborg Society, 1961; (2) also in *Posthumous Theological Works.* Second edition. West Chester, Penna.: Swedenborg Foundation, 1996. A recent abridged edition is *The Last Judgment in Retrospect.* Translated and edited by George F. Dole. West Chester, Penna.: Swedenborg Foundation, 1997.

Doctrina Novae Hierosolymae de Domino, de Scriptura Sacra, de Fide; Doctrina Vitae pro Nova Hierosolyma ex praeceptis decalogi. Amsterdam, 1763. English translation, *The Four Doctrines.* Translated and edited by John F. Potts. Second edition. West Chester, Penna.: Swedenborg Foundation, 1997.

Sapientia Angelica de Divino Amore et de Divina Sapientia. Amsterdam, 1763. English translation, *Divine Love and Wisdom.* Translated by John C. Ager. Second Edition. West Chester, Penna.: Swedenborg Foundation, 1996. Also *Divine Love and Wisdom.* Translated by George F. Dole. West Chester, Penna.: Swedenborg Foundation, 1985.

Sapientia Angelica de Divina Providentia. Amsterdam, 1764. English translations, *Divine Providence.* Translated by William F. Wunsch. Second Edition. West Chester, Penna.: Swedenborg Foundation, 1996.

Apocalypsis Revelata in qua deteguntur arcana quae ibi praedicta sunt, et hactenus recondita latuerunt. Amsterdam, 1766. English translation, *The Apocalypse Revealed.* Translated by John Whitehead. Two volumes. Second Edition. West Chester, Penna.: Swedenborg Foundation: 1997.

Deliciae Sapientiae de Amore Conjugiali, post quas sequuntur voluptates insaniae de amore scortatorio. Amsterdam, 1768. English translation, *Conjugial Love.* Translated by Samuel S. Warren. Revised by Louis Tafel. Second edition. West Chester, Penna.: Swedenborg Foundation, 1998. Also *Love in Marriage.* Translated by David F. Gladish. West Chester, Penna.: Swedenborg Foundation, 1992.

Summaria expositio doctrinae Novae Ecclesiae, quae per Novam Hierosolymam in Apocalypsi intelligitur. Amsterdam, 1769. English translation, *A Brief Exposition of the Doctrine of the New Church.* In *Miscellaneous Theological Works.* Translated by John Whitehead. Second edition. West Chester, Penna.: Swedenborg Foundation, 1996.

De Commercio Animae et Corporis. London, 1769. English translation, *The Interaction of the Soul and the Body.* In *Miscellaneous Theological Works.* Translated by John Whitehead. Second Edition. West Chester, Penna.:

Swedenborg Foundation, 1996. Also *Emanuel Swedenborg: The Universal Human and Soul-Body Interaction.* Translated by George F. Dole. New York: Paulist Press, 1984.

Vera Christiana Religio, continens universam theologiam Novae Ecclesiae. Amsterdam, 1771. English translation, *The True Christian Religion.* Translated by John C. Ager. Second Edition. West Chester, Penna.: Swedenborg Foundation, 1996.

Emanuel Swedenborg's Journal of Dreams. Stockholm 1859. Translated by J.J.G. Wilkinson. Commentary by Wilson Van Dusen. New York: Swedenborg Foundation, 1986. *Swedenborgs Drömbok: Gladjen och det Stora Kvalet.* Edited by Lars Bergquist. Stockholm: Norstedts, 1988.

Secondary Sources

The following list is a severely restricted selection of the enormous bulk of Swedenborg criticism. It consists mainly of books and papers that I have found useful in preparing my book, although not all of them are found in the notes.

Aarsleff. "Leibniz on Locke on Language." *American Philosophical Quarterly* I (1964): 1–25.

Acton, Alfred, trans. and ed. *Letters and Memorials of Emanuel Swedenborg.* Bryn Athyn: Swedenborg Scientific Association, 1948–1955. A much more complete collection of Swedenborg's correspondence than the one found in Tafel's *Documents*, with informative annotations by the editor, one of the very best of New Church scholars.

______. *Psychologica, being Notes and Observations on Christian Wolff's Psychologia Empirica by Emanuel Swedenborg.* Philadelphia: Swedenborg Scientific Association, 1923. A Latin–English edition with an important preface by the editor.

______. *A Philosopher's Note Book. Excerpts from Philosophical Writers and from the Sacred Scriptures on a Variety of Philosophical Subjects; together with some Reflections, and sundry Notes and Memoranda by Emanuel Swedenborg.* Philadelphia: Swedenborg Scientific Association, 1931. This is an annotated translation of ms.cod. 36-110 in the Royal Swedish Academy of Science, which is invaluable to anyone interested in Swedenborg's philosophical erudition around 1740.

———. *An Introduction to The Word Explained. A Study of the Means by which Swedenborg the Scientist and Philosopher became the Theologian and Revelator.* Bryn Athyn, Penna.: Academy of the New Church, 1927. A detailed study of Swedenborg's literary life up to the first volume of *Arcana Coelestia.*

Antón Pacheco, José Antonio. *Un libro sobre Swedenborg.* Seville, Spain: Publicaciones de la Universidad de Sevilla, 1991.

Arbman, Ernst. *Ecstacy or Religious Trance.* Stockholm: Svenska Bokförlaget, 1963.

Arrhenius, Svante. "Emanuel Swedenborg as a Cosmologist." In *Opera quaedam aut inedita aut obsoleta de rebus naturalibus.* Vol. II. Stockholm: The Royal Swedish Academy of Science, 1908. A sober evaluation of Swedenborg's cosmological position by the famous Swedish Nobel laureate.

Auerbach, Erich. *Mimesis: The Representation of Reality in Western Literature.* Translated by William Trask. New York: Doubleday Anchor Books, 1957. One of the classics of twentieth-century criticism.

Bellin, H .F. and Ruhl, D., eds. *Blake and Swedenborg: Opposition is True Friendship.* New York: Swedenborg Foundation, 1985.

Benz, Ernst. " Swedenborg und Lavater."*Zeitschrift für Kirchengeschichte* 57 (1938): 153–216.

———. *Swedenborg in Deutschland: F.C. Oetingers und Immanuel Kants Auseinander-setzung mit der Person und Lehre Emanuel Swedenborgs nach neuen Quellen bearbeitet* Frankfurt am Main: Vittorio Klostermann, 1947. A thorough study of Swedenborg's early reception in Germany.

———. *Emanuel Swedenborg: Naturforscher und Seher.* München: Hermann Rinn, 1948.This is an extensive biography, which is especially interesting from the point of view of ecclesiastical history. Unfortunately, the author gives no references to Swedenborg research, and there are mistakes regarding Swedenborg's scientific contributions and position. An English translation is scheduled for publication in 2001.

Bergquist, Lars. *Swedenborgs Drömbok. Glädjen och det stora kvalet.* Stockholm: Norstedts, 1988. An annotated edition (in Swedish) of the *Journal of Dreams* with many valuable commentaries and interesting observations. An English translation will soon be available.

———. *Biblioteket i lusthuset.* Stockholm: Natur och Kultur, 1996. Ten essays on Swedenborg (in Swedish) and a Swedish translation of a Latin dissertation published in 1787, which attacks Swedenborg as a Neo-Platonic visionary.

Bergmann, Horst, and Eberhard Zwink. *Emanuel Swedenborg, 1688–1772: Naturforscher und Kundiger der Überwelt.* Stuttgart, Germany: Württembergische Landesbibliothek, 1988. A collection of essays by European scholars in honor of the tricentennial of Swedenborg's birth.

Block, Marguerite Beck, *The New Church in the New World. A Study of Swedenborgianism in America.* 1932; reprinted New York: Swedenborg Publishing Association, 1968. The standard book on American Swedenborgianism with a very valuable bibliography.

———. "Scientist into Seer: The Psychological Problem Presented by Swedenborg." *The Review of Religion* (May 1938): 412–432. An interesting discussion of Swedenborg's religious crisis in 1744–1745 and the insanity diagnoses.

Brandell, Gunnar. *Strindbergs Infernokris.* Stockholm: Bonniers, 1950.

Brock, Erland, ed. *Swedenborg and His Influence.* Bryn Athyn, Penna.: Academy of the New Church, 1988. A collection of essays by Swedenborgian scholars in honor of the tricentennial of Swedenborg's birth.

Calatrello, R. L. *The Basic Philosophy of Emanuel Swedenborg, with Implications for Western Education.* Diss. University of Southern California, 1966. Microfilm.

Christensson, Jakob. *Lyckoriket. Studier i svensk upplysning.* Stockholm: Atlantis, 1996. An informative (and voluminous) dissertation about, among other things, the early reception of Swedenborg's works in Sweden. In Swedish but with a summary in English.

Couturat, L., and L. Leau. *Histoire de la langue universelle.* Paris: Librairie Hachette, 1903.

Croce, Paul J. "A Scientific Spiritualism: The Elder James's Adaptation of Emanuel Swedenborg." In *Swedenborg and His Influence*, ed. Erland Brock. Bryn Athyn, Penna.: The Academy of the New Church, 1988.

Darnton, Robert. *Mesmerism and the End of the Enlightenment in France.* Cambridge, Mass.: Harvard University Press, 1968.

———. *The Literary Underground of the Old Regime.* Cambridge, Mass.: Harvard University Press,1982.

Dingwall, E. J. *Very Peculiar People: Portrait Studies in the Queer, the Abnormal and the Uncanny.* New York: University Books, 1962. Contains an interesting and well-informed portrait of Swedenborg from the point of view of parapsychology.

Eby, S.C. *The Story of the Swedenborg Manuscripts.* New York: New-Church Press, 1926.

Edel, Leon. *Henry James.* Two volumes. London: Rupert Hart-Davis, 1953, 1962.

Edmisten, Leonard Martin. *Coleridge's Commentary on Swedenborg.* Dissertation, University of Missouri, 1954. Microfilm.

Emerson, Ralph Waldo. *Representative Men.* London: Gibbings and Company, 1901.

Eneström, Gustaf. "Emanuel Svedenborg såsom matematiker." In *Transactions of the Royal Swedish Academy of Science,* vol. XV. Stockholm: 1890. An expert's evaluation of Swedenborg as a mathematician (in Swedish).

Florschütz, Gottlieb. *Swedenborg and Kant: Emanuel Swedenborg's Mystical View of Humankind and the Dual Nature of Humankind in Immanuel Kant.* Translated by George F. Dole. West Chester, Penna.: Swedenborg Foundation, 1993.

Frängsmyr, Tore. *Geologi och skapelsetro.* Uppsala: Almqvist & Wiksell, 1969. An exhaustive study with a summary in English (Geology and the Doctrine of Creation) of, among others, Swedenborg's writings on cosmological and geological issues.

———. *Wolffianismens genombrott i Uppsala* (The Emergence of Wolffianism at Uppsala: University Philosophy from 1720 up to the Middle of the Eighteenth Century). Uppsala: Almqvist & Wiksell,1972. Summary in English.

Geymüller, Henri de. *Swedenborg et les phénomènes psychiques.* Paris, 1934. German translation, P. Sakmann. *Swedenborg und die übersinnliche Welt.* Stuttgart: Deutsche Verlags-Anstalt, 1936. A survey of Swedenborg's experiences as a seer, which is of some interest but must be read very critically.

Griffith, Freda G. *The Swedenborg Society 1810–1960.* London: Swedenborg Society, 1960.

Gruhle, Hans W. "Swedenborgs Träume. Ein Beitrag zur Phänomenologie seiner Mystik." *Psychologische Forschung* 54 (1924). An interesting analysis of the *Journal of Dreams* and a very sensible evaluation of the insanity problem.

Hallengren, Anders. *The Code of Concord: Emerson's Search for Universal Laws.* Stockholm: Almqvist & Wiksell International, 1994. An impressive dissertation that studies, among other things, the influence of Swedenborg on Emerson's thinking.

———. *Tingens tydning.* Stockholm: Åsak, 1997.

———. *Öarna under vinden.* Stockholm: Åsak, 1997.

———. *Gallery of Mirrors: Reflections of Swedenborgian Thought.* West Chester, Penna.: Swedenborg Foundation, 1998. A collection of essays that examine the implications of Swedenborg's influence.

Hazard, Paul. *La crise de la conscience européenne.* Paris: Boivin & Cie, 1935. A classic study of the Enlightenment in Europe.

Heinrichs, Michael. *Emanuel Swedenborg in Deutschland. Eine kritische Darstellung der Rezeption des schwedischen Visionärs im 18. und 19. Jahrhundert.* Frankfurt: Europäische Hochschulschriften Reihe XX, vol. 47, 1979.

Hitschmann, Eduard. "Swedenborgs Paranoia." In *Zentralblatt für Psychoanalyse.* 1912. A psycho-analytic interpretation of Swedenborg.

Hjern, Olle. "Carl Jonas Almqvist: Great Poet and Swedenborgian Heretic." In *Swedenborg and His Influence.* Bryn Athyn, Penna.: Academy of the New Church, 1988.

Hobart, Nathaniel. *Life of Emanuel Swedenborg with Some Account of His Writings.* Third edition prepared by Benjamin Worcester. New York: John Allen, 1850. An old biography, still worth reading.

Hoover, Dwight D. "The Influence of Swedenborg on the Religious Ideas of Henry James, Senior." In *Swedenborg and His Influence.* Edited by Erland Brock, et al. Bryn Athyn, Penna.: Academy of the New Church Press, 1988.

Horn, Friedemann, *Schelling and Swedenborg: Mysticism and German Idealism.* Translated by George F. Dole. West Chester, Penna.: Swedenborg Foundation, 1997.

Häll, Jan. *I Swedenborgs labyrint. Studier i de gustavianska swedenborgarnas liv och tänkande.* Stockholm: Atlantis, 1995. An informative dissertation on the earliest Swedenborg followers in Sweden, with a summary in English.

Iversen, Erik. *The Myth of Egypt and its Hieroglyphs in European Tradition.* Copenhagen: Gad, 1961.

Jalabert, Jacques. *Le Dieu de Leibniz.* Paris: Presses Universitaires de France, 1960.

Jaspers, Karl. *Strindberg und van Gogh: Versuch einer pathographischen Analyse unter vergleichende Heranziehung von Swedenborg und Hölderlin.* New edition. Bremen: J. Storm, 1949. The famous philosopher regards Swedenborg as a typical example of schizophrenia.

Johannisson, Karin. *Magnetisörernas tid: Den animala magnetismen i Sverige.* Dissertation, Uppsala University, 1974. A most interesting study of the Swedish practitioners of animal magnetism in the late-eighteenth century, some of whom were followers of Swedenborg. Summary in English.

Jonsson, Inge. *Swedenborgs skapelsedrama De cultu et amore Dei* (Swedenborg's Drama of Creation The Worship and Love of God. Stockholm: Natur och Kultur, 1961. A dissertation with a summary in English.

______. *Swedenborgs korrespondenslära* (Swedenborg's Doctrine of Correspondence). Stockholm: Almqvist & Wiksell International, 1969. Summary in English.

Keller, Helen. *Light in My Darkness.* Edited by Ray Silverman. West Chester, Penna.: Swedenborg Foundation, 1994. A revised edition of Keller's "spiritual autobiography" *My Religion*, originally published in 1927.

Kirven, Robert. *Emanuel Swedenborg and the Revolt against Deism.* Dissertation, Brandeis University, Waltham, 1965. Microfilm.

Kleen, E. A. G. *Swedenborg, en lefnadsskildring.* Two volumes. Stockholm: Sandbergs, 1917, 1920. A violently polemical Swedish biography which, however, has some value in the analysis of Swedenborg's contributions to science.

Kübler-Ross, Elisabeth. *The Wheel of Life. A Memoir of Living and Dying.* New York: Station Hill Press, 1997.

Lagercrantz, Olof. *Dikten om livet på den andra sidan. En bok om Emanuel Swedenborg.* Stockholm: Wahlström & Widstrand, 1996. German translation by Angelika Gundlach, *Von Leben auf der anderen Seite* (Frankfurt au Main: Suhrkamp Verlag, 1997).

Lamm, Martin. *Swedenborg. En studie öfver hans utveckling till mystiker och andeskådare* 1915; reprinted Stockholm: Hammarström & Åberg, 1987. German translation by Ilse Meyer-Lüne (Leipzig, 1922). French translation by E. Söderlindh with a preface by Paul Valéry (Paris, 1936). An English translation is scheduled for publication in 2000. This is still the most important monograph on Swedenborg.

Larsen, Robin, et al, eds. *Emanuel Swedenborg: A Continuing Vision.* New York: Swedenborg Foundation, 1988. A pictorial biography and collection of essays by prominent Swedenborgian scholars and international figures.

Lenhammar, Harry. *Tolerans och bekännelsetvång. Studier i den svenska swedenborgianismen 1765–1795.* Uppsala: Acta Universitatis Upsaliensis, 1966. A most valuable dissertation in ecclesiastical history about the early Swedenborg disciples in Sweden, with a summary in English.

Lindborg, Rolf. *Descartes i Uppsala. Striderna om "nya filosofien" 1663–1689.* Uppsala: Almqvist & Wiksell, 1965. A most important dissertation about the conflicts caused by Cartesianism in Sweden,with a summary in English.

Lindqvist, Svante. *Technology on Trial: The Introduction of Steam Power Technology into Sweden, 1715–1736.* Stockholm: Almqvist & Wiksell International, 1984.

Lubac, Henri de. *Exégèse médiévale. Les quatre sens de l'Ecriture.* Two volumes. Paris: Aubier, 1959, 1964.

Mazzeo, Joseph Anthony. *Renaissance and Seventeenth Century Studies.* New York: Columbia University Press, 1964.

Michaud, Guy. *La doctrine symboliste.* Paris: Librairie Nizet, 1947.

———. *Message poétique du symbolisme.* Paris: Librairie Nizet, 1951.

Moreau, Joseph. *L'Univers Leibnizien.* Paris: Emanuel Vitte, 1956.

Pasquale, Elisabetta Duina. "Blake/Swedenborg." Ph.D. dissertation, University of Milan, 1992/1993. In Italian.

Potts, J. F. *Swedenborg Concordance.* 6 volumes. London: Swedenborg Society, 1888–1902.

Praz, Mario. *Studies in the Seventeenth Century Imagery.* London: Warburg Institute, 1939.

Ramström, Martin. "Swedenborg on the Cerebral Cortex as the Seat of Psychical Activity." In *Transactions of the International Swedenborg Congress 1910.* Third edition. London: Swedenborg Society, 1912, 56–70. A summary of two important papers in Swedish, in which the author investigates Swedenborg's sources for his brain anatomy and physiology.

Roger, Jacques. *Les sciences de la vie dans la pensée francaise du XVIIIe siècle.* Paris: Armand Colin, 1963.

Roos, Jacques. *Aspects littéraires du mysticisme philosophique et l'influence de Boehme et de Swedenborg au début du romantisme: William Blake, Novalis, Ballanche.* Strasbourg: P.H. Heitz, 1951. A rewarding study of Swedenborg's influence on Romanticism in its early stages.

Russell, Bertrand. *A Critical Exposition of the Philosophy of Leibniz.* Second edition. London: George Allen & Unwin, Ltd., 1949.

Schlieper, Hans. *Emanuel Swedenborgs System der Naturphilosophie, besonders in seiner Beziehung zu Goethe-Herderschen Anschauungen.* Berlin: Gustave Schade, 1901. A small but valuable doctoral thesis on Swedenborg's philosophy of nature.

Schöne, Albrecht. *Emblematik und Drama im Zeitalter des Barock.* Munich: C.H. Beck'sche Verlagsbuchhandlung, 1964.

Sigstedt, Cyriel Odhner. *The Swedenborg Epic: The Life and Works of Emanuel Swedenborg.* 1952; reprinted London: Swedenborg Society, 1981. A comprehensive biography that bears the stamp of the author's Swedenborgian background but is almost free from such exaggerations and historical mistakes that disfigure much of older New-Church scholarship. Sigstedt carefully accounts for the result of non-Swedenborgian research, including Martin Lamm's epoch-making study, and her book is of great value.

Silver, R. K. "The Spiritual Kingdom in America": The Influence of Emanuel Swedenborg on American Society and Culture, 1815–1860. Ph.D. dissertation, Stanford University, 1983. Microfilm.

Sjödén, Karl-Erik. *Swedenborg en France.* Stockholm: Almqvist & Wiksell International, 1985. Much previously unknown material concerning the Swedenborg's reception in eighteenth- and nineteenth-century France.

Skelton, Robin, and Ann Saddlemeyer, eds. *The World of W.B. Yeats.* Victoria, B.C., Canada: University of Victoria, 1965.

Stockenström, Göran. "'The Great Chaos and the Infinite Order': The Spiritual Journeys of Swedenborg and Strindberg." In *Swedenborg and His Influence.* Bryn Athyn, Penna.: Academy of the New Church Press, 1988.

Stroh, Alfred H. "Analysis and Review of *The Worship and Love of God.*" *The New Philosophy* 5 (1902): 33–75. A very useful study of Swedenborg's drama of creation by a prominent Swedenborgian scholar.

______. "Swedenborg's Early Life, Scientific Works, and Philosophy." *The New-Church Magazine* XXXIV (1915): 172–179, 204–210, 262–269, 354–359, 396–403, 440–445, 490–498, 540–547; vol. XXXV (1916): 6–14. An interesting analysis of Swedenborg's writings up to *Principia,* based on the author's strenuous studies of archives in Sweden.

Tafel, Rudolph L. *Documents concerning the Life and Character of Emanuel Swedenborg.* Collected, translated and annotated by R .L. Tafel. Two volumes (volume two in two parts). London: Swedenborg Society, 1875–1877. A very useful tool for all Swedenborg research but unfortunately filled with inadvertencies, not to say grave mistakes.

Tice, David. *Tekel: The Credentials and Teachings of Emanuel Swedenborg Examined.* Cincinnati, Ohio: Jennings and Pye, 1901. The standard criticism of Swedenborg in English.

Toksvig, Signe. *Emanuel Swedenborg, Scientist and Mystic.* 1948, reprinted New York: Swedenborg Foundation, 1983. A biography written with journalistic brio, of special interest from the point of view of parapsychology; however, it should be used with some caution, since many interpretations seem to be a bit too imaginative, for example, Toksvig's evaluation of Swedenborg as a scientist.

Trobridge, George. *Swedenborg, Life and Teaching.* Fifth edition. Revised by R. H. Tafel Sr. and R. H. Tafel Jr. New York: Swedenborg Foundation, 1992. The semi-official New Church biography.

Unterecker, John, ed. *Yeats. A Collection of Critical Essays.* Englewood Cliffs, New Jersey: Prentice-Hall, 1963.

Viatte, Auguste. *Les Sources occultes du romantisme.* Two volumes. Paris: Librairie Ancienne Honoré Champion, 1928.

Warren, S. M. *A Compendium of the Theological Writings of Emanuel Swedenborg.* 1875; reprinted New York, Swedenborg Foundation, 1979. A useful reference book for those interested in Swedenborg's theology.

Wilkinson, J. J .G. *Emanuel Swedenborg, a Biographical Sketch.* Second edition. London: James Speirs, 1886. The author was an eminent expert on Swedenborg's scientific writings, especially in anatomy and physiology, and the book is still worth reading, in spite of its many factual errors.

Wilkinson, Lynn R. *The Dream of an Absolute Language. Emanuel Swedenborg and French Literary Culture.* Albany: State University of New York Press, 1996.

Williams-Hogan, Jane. *A New Church in a Disenchanted World: A Study of the Formation and Development of the General Conference of the New Church in Great Britain.* Dissertation, University of Pennsylvania (Philadelphia), 1985. Microfilm.

Woofenden, William Ross. *Swedenborg Researcher's Manual.* Bryn Athyn, Penna.: The Swedenborg Scientific Association, 1988. A most useful introduction to Swedenborg scholarship.

Worcester, Benjamin. *The Life and Mission of Emanuel Swedenborg.* Philadelphia: J. B. Lippincott and Co., 1883. An extensive New Church biography.

Yeats, William Butler. *Autobiographies.* London: Macmillan & Co., Ltd, 1955.

Index

Acton, Alfred, 51
Adolf Frederick, crown prince of Sweden, 123
Africans, 170–171
Alciati, Andreas, 112
Almquist, Carl Jonas Love, 208
Ariosto, Ludovico, 154
Aristippus, 169
Aristotle, 33, 58, 61, 70, 71, 74, 76, 77, 82, 111, 119, 170
Augustine, 80, 81, 82, 86, 119, 179, 190–191

Baglivi, G., 40
Balzac, Honoré de, 7, 203–204, 207, 210
Baron, Robert, 75
Baudelaire, Charles, 7, 205–207, 210
Behm, Albrecht, 9
Behm, Sara, 9
Benzelius, Erik, Jr., 10, 12, 14, 19, 178
Bergia, Sara, 10
Bilfinger, Georg, 119
Blake, William, 7, 200, 202
Block, Marguerite Beck, 199
Boehme Jacob, 148, 175–176, 202
Boerhaave, Herman, 45, 65
Borelli, Giovanni Alfonso, 40
Borges, Jorge Luis, 7, 210
Boyle, Robert, 121
Breton, André, 207
Browne, Sir Thomas, 113
Buffon, George Louis Leclerc, 26, 28
Burnet, Thomas, 27–28, 139

Cambridge Platonism, 58, 81
Champollion, Jean François, 112
Charles XII, king of Sweden, 12, 13, 19, 64
Coleridge, Samuel Taylor, 28
Collegium Curiosorum, 10, 14
Comenius, Amos, 113, 114–115
Copernicus, Nicolaus, 35
Cortesius, G. B., 40
Cudworth, Ralph, 58, 60

Dante Alighieri, 19, 118
Darnton, Robert, 204
Democritus, 170
Descartes, René, 4–5, 25–26, 31, 33, 35–36, 41, 42, 46, 54, 57, 58, 61, 70, 81, 89, 94, 119, 147
Diderot, Denis, 43
Dionysius the Areopagite, 190
Dipple, Johan Conrad, 176
Donne, John, 3
Dupleix, Scipion, 75

Eckhart, Meister, 190
Ekelöf, Gunnar, 210
Ekelund, Vilhelm, 7, 202, 210
Ekman, Kerstin, 210
Eliot, T. S., 3, 7
Elvius, Petrus, 22

Index

emblematics, 107, 112–115, 164
Emerson, Ralph Waldo, 21, 201–202, 210
Enlightenment, the, 5, 6, 16, 24, 43, 197
Ernesti, Johann August, 196
Exegetic-Philanthropic Society, 197, 199

Flamsteed, John, 11
Fontenelle, Bernard, 154
Frängsmyr, Tore, 64

Galilei, Galileo, 14
Geulincx, Arnold, 70
Gjörwell, Carl Christopher, 196
Goethe, Johann Wolfgang, 6, 198, 202
Görres, Joseph, 198
Grotius, Hugo, 119
Gustav III, king of Sweden, 197
Gyllenborg, Elisabeth, 167
Gyllensten, Lars, 210

Habermas, Jûrgen, 161
Hallengren, Anders, 187
Halley, Edmond, 11
harmonia praestabilitia (Leibniz–Wolff), 70, 72, 84–88, 89, 98. *See also* Swedenborg, *harmonia constabilita*
Harvey, William, 35–36
Heister, L., 5, 47
Helmont, John Baptiste, 58
Helvétius, Claude Adrien, 94
Heraclitus, 170
Herder, Johann Gottfried, 112, 148
Homer, 23, 24
Hooke, Robert, 121
Höpken, Anders Johan, 196
Horace, 114
Horapollo, 112
Hugo, Victor, 7, 113, 204

Illuminism, 203

James, Henry, Jr., 201
James, Henry, Sr., 200–201
James, William, 201

Kabbalah, 148, 175, 178, 179
Kant, Immanuel, 20, 26, 28, 133, 198
Kellgren, Johan Henrik, 197
Kepler, Johannes, 59
Kleen, Emil A. G., 132
Klemming, Gustaf, 208

Lagercrantz, Olof, 210
La Mettrie, Julien Offray, 5, 43
Lamartine, Alphonse, 204
Lamm, Martin, 6, 78, 80, 81, 124, 132, 145, 191
Langbehn, Julius, 198
Laplace, Pierre Simon, 26, 28
Lavater, Johann Caspar, 198
Law, William, 175
Leeuwenhoek, Anthony, 36–37, 47, 51, 55, 62, 65, 66, 67, 71, 74, 78, 88
Leibniz, Gottfried Wilhelm, 5, 6, 10, 30, 38–39, 59, 60–64, 70, 71, 72, 74, 77, 81, 84, 85, 86, 87, 88, 91, 94–96, 98, 100, 107, 119, 151, 153, 187, 190
Linné, Carl von, 15, 35, 123
Locke, John, 59, 77, 78–79, 102
Louisa Ulrika, queen of Sweden, 132–133
Lubac, Henri de, 178
Lucian, 154
Lull, Raimon, 94

Malebranche, Nicolaus, 5, 38, 59, 70, 86, 104, 119, 120, 121, 152, 153, 190, 191
Malpighi, Marcello, 38, 47, 56, 61, 62, 64, 78
Marco Polo, 187
Michaud, Guy, 205

Milosz, Czeslaw, 210
Milton, John, 138, 139
More, Henry, 58
Morgagni, Giovanni Battista, 65

Neo-Platonism, 80–81, 111, 190
Neumann, Caspar, 116, 147
New Church, the, 196, 199–200, 210
Newton, Isaac, 5, 14, 22, 24–25, 26, 56, 59, 90, 100, 121
Nicholas of Lyra, 179–180
Nietzsche, Friedrich, 106, 202

Oetinger, Friedrich Christoph, 143, 160, 196, 198
Origen, 179, 180, 190
Ovid, 26, 28, 118, 139
ovism, theory of, 36–39. *See also* world egg.

Paracelsus, 58
Pernety, Abbé A. J., 203
Philo, 178
Picinelli, Philippus, 114
Plato, 54, 89, 113, 119, 170
Plotinus, 81, 190, 191
Polhem, Christopher, 10, 12–13, 15, 19, 24, 39
Polhem, Emerentia, 13
Pope, Alexander, 24
Praz, Mario, 114
Prester, John, 171, 186–187
Pythagoras, 73, 113, 169

Réaumur, René Antoine, 60
Ridley, H., 47, 65
Roger, Jacques, 59–60
Romanticism, 6, 63, 64, 86, 112
Rosseau, Jean-Jacques, 112
Rydelius, Andreas, 119

Sade, Donatien Alphonse François de, 132
Sand, George, 204
Schelling, Friedrich, 198
Scholasticism, 4, 70, 74, 115
Schönström, Peter, 187
Schubert, G. H., 198
Societas Literaria et Scientiarum, 14
Socrates, 169
solifidianism, 87, 185
Spinoza, Barusch, 6, 70
Stagnelius, Erik Johan, 7
Steno, Nicolaus, 5, 121
Strahlenberg, Philip Johan, 187
Strindberg, August, 7, 207–210
Stroh, Alfred H., 28
Sturm, Johann Christopher, 58
Swammerdam, Jan, 5, 38, 47, 50, 55, 62, 72, 121
Swedberg, Jesper, 9–10, 15, 130
Swedenborg, Emanuel
  anima, 74, 76–77, 79, 81, 82, 85, 87, 94, 104, 135, 136, 144, 189. *See also* Swedenborg, soul.
  animus, 76, 77, 79, 80, 103, 137, 180
  brain physiology, 44–45, 49–54, 55, 84, 97, 104, 105
  Christology, 138–139, 189. *See also* Swedenborg, theology.
  cosmology, 22–29
  doctrine of correspondence, 7, 70, 80, 81, 87, 88, 93–121, 128, 146, 148, 160, 165, 168, 173, 177, 183, 184–188, 189, 202, 206, 207
  doctrine of forms, 88–91, 145, 148, 173
  doctrine of series and degrees, 69–76, 83, 97, 99, 115, 128, 177, 190
  evil, 191
  exegesis, 175–193

Index

harmonia constabilita, 72, 84, 87, 117. *See also harmonia praestabilitia.*
imaginatio, 103
influence, 195–210
intellectus purus, 104
longitude problem, 11–12
marriages in heaven, 164–167
mathematical point, 25, 29–32, 52
mathematics, interest in, 11, 12, 14, 25, 94, 96–97
Maximus Homo (Grand Man), 54, 151–154
mens intellectualis, 76
mens rationalis, 76–77, 79, 80, 82, 103, 137
mental state, 130–133
planetary theory, 26, 27, 28
prince of the world, 136, 138–139
psychological system, 76–84, 97
publication of manuscripts, 13, 47–48, 199
regeneration, 192
sense physiology (nerve tremors), 39–41, 43, 47
series of particles and elements, 30–32, 41, 65–66
soul, 42–46, 47, 61, 69–70, 73, 74–75, 81, 93, 105, 137. *See also* Swedenborg, *anima.*
spirits from other planets, 154–155
spiritual speech, 145–149
spiritual world, 143–173
theology, 188–199
travel journals, 15–16, 47, 123–124, 159
vis formatrix (formative force), 56–65, 82
visions, 125, 133, 140, 144–145
will, 87, 190–191
WRITINGS OF,
Adversaria in libros Veteris Testamenti (The Word of the Old Testament Explained),177, 180, 181
Apocalypsis Revelata (Apocalypse Revealed), 186
Arcana Coelestia (Heavenly Secrets), 24, 54, 85, 111, 114, 142, 143–147, 149, 159, 160, 166, 173, 176, 181, 182–188, 185, 198, 208
Clavis Hieroglyphica, 106–112, 116–118, 119, 121, 158, 180
Daedalus Hyperboreus, 12–13, 39, 84
De Anima (On Spirit), 76, 80, 87, 103, 105, 110, 117, 118, 136, 137, 145
De Commercio Animae et Corporis (Interaction of Soul and Body), 69–70
De Coelo et Ejus Mirabilibus (Heaven and Hell), 145, 158–160, 200
De Cultu et Amore dei (Worship and Love of God), 26, 28, 47, 48,. 53, 57, 83, 104, 120, 121, 129, 133–139, 140, 157–158, 160, 181, 191
De Equo Albo (The White Horse), 168
De Infinito (On the Infinite), 32–34, 42, 64, 75, 83, 153, 188, 189
De Ultimo Judicio (The Last Judgment), 153–157
Delicia Sapientiae de Amore Conjugali (Conjugial Love), 149, 159–172, 186, 191
Diarium Spirituale (Spiritual Diary), 18, 105, 127, 130, 131, 141, 144, 146, 148, 149, 173, 176, 178

Index

Doctrina Novae Hierosolymae de Scriptura Sacra (The New Jerusalem in Holy Scripture), 184
Journal of Dreams, 16–19, 121, 123–130, 137, 144, 165, 209
Miscellanea observata circa res naturales, 30
Oeconomia Regni Animalis (Economy of the Animal Kingdom), 48 51, 54, 56, 58, 60, 65, 70, 72, 73, 75, 79, 80, 86, 88, 97, 102, 103, 117, 120, 128, 129, 130, 131, 135, 137, 144, 148, 163, 177, 190
Opera Philosophica et Mineralia, 14, 21, 22
Philosophia Corpuscularis, 65
Principia Rerum Naturalium, 23, 24, 27, 30, 31, 41, 45, 46, 52, 64, 65, 98, 135, 192
Prodromus Principorum Rerum Naturalium, 30
Regnum Animale (The Animal Kingdom), 17, 48, 76, 81, 109, 110, 123, 125, 127, 130
Sapientia Angelica de Divina Providentia (Divine Providence), 200
Sapientia Angelica de Divino Amore (Divine Love and Wisdom), 49, 200
Vera Christiana Religio (True Christian Religion), 20, 157, 160, 171, 184, 186, 188, 192, 209

Tafel, Rudolph L., 8, 27, 50
Tilas, Daniel, 196
Tindall, William York, 202
Tournefort, Joseph Pitton de, 64

universal language, 4, 55–56, 93–106, 177

Valéry, Paul, 7
Vallisneri, Antonio, 64
Verheyen, Ph., 47
Vesalius, Andreas, 35
Vico, Giambattista, 112
Vieussens, Raymond, 40, 47, 84
Voltaire, 5, 25, 58–59, 64, 79

Walser, Martin, 199
Wilkins, John, 94, 110
Willis, Thomas, 40
Winslow, Jacob Benignus, 5, 47, 65, 78, 80
Wolf, Christopher, 14
Wolff, Christian, 30, 46, 64–65, 70, 71, 72, 74, 75, 87, 88, 89, 90, 91, 98, 99–100, 102, 108, 115, 116, 119, 147, 185
Woodward, John, 11
world egg (*ovum mundanum*), 26, 27, 28, 29, 134–135, 139

Xenophon, 169

Yeats, William Butler, 7, 202–203

Zenon, 30